MASONRY

Beyond The Light

by William Schnoebelen

Published by CHICK PUBLICATIONS
P. O. Box 662, Chino, CA 91708-0662
Printed in the United States of America

INTERNATIONAL DISTRIBUTORS

Comic-Traktate-Versand
Postfach 3009
5632 Wermelskirchen 3
West Germany

Christ is the Answer, Inc.
Box 5167, Station A
Toronto, Ontario M5W 1N5, Canada
(416) 699-7800

Chick Publications Distributor
P. O. Box 50096
Porirua, Wellington, New Zealand

Penfold Book & Bible House
P. O. Box 26
Bicester, Oxon, England OX6 8PB
Tel: 0869-249574

Evangelistic Literature Enterprise
P. O. Box 10
Strathpine, Q'ld., Australia 4500
Phone: (07) 205-7100

ISBN:09-937958-39-5

2nd Printing

193/A

DEDICATION

To Pastor Don and LaVena Allen, who were patient and loving in their discipleship of us, and who taught us much of what we know about serving Jesus Christ, and about the ministry of deliverance.

Contents

PART 2

The History of Masonry

This book has been a tremendous effort, and I would like to thank the many people who prayed for its completion. The Deceiver does not like books written on this difficult and controversial subject, and the spiritual warfare has been intense.

I acknowledge humbly the prayer support of the people (too numerous to mention) behind this book, and pray it will bring glory and honor to the name of my "Worshipful Master," Jesus Christ.

Some extraordinary people were indispensable in putting this book together. I would especially like to thank my wife, Sharon, for her loving support and patience with her "writer husband," as well as her thoughtful insights and immense help with the typing and editing.

I would like to thank Ed Decker, who was both an important inspiration and a helpful editor and critic during the formative stages of this project.

I need to also mention the helpful insights, wisdom and research assistance from Mick Oxley, Aron Rush and Jim Zilonka.

The purpose of this book is to *speak the truth in love* (Ephesians 4:15) that many might be brought from darkness into light!

William J. Schnoebelen

...and the spirit lifted me up between the earth and the heaven, and brought me in the visions of God to Jerusalem, to the door of the inner gate that looketh toward the north; where was the seat of the image of jealousy, which provoketh to jealousy.

And, behold, the glory of the God of Israel was there, according to the vision that I saw in the plain.

Then said he unto me, Son of man, lift up thine eyes now the way toward the north. So I lifted up mine eyes the way toward the north, and behold northward at the gate of the altar this image of jealousy in the entry.

He said furthermore unto me, Son of man, seest thou what they do? even the great abominations that the house of Israel committeth here, that I should go far off from my sanctuary? but turn thee yet again, and thou shalt see greater abominations.

Ezekiel 8:3-6

INTRODUCTION

THE LAMB OR THE LAMBSKIN?

The hot noonday sun streamed down as I stepped from my car. Though it was a bright Iowa summer day, the light in my own heart was even brighter! As I walked across the street to the Masonic Lodge in my town, there was a spring in my step which nothing but Jesus Christ could bring!

God was in His heaven, and all seemed right with the world. I had given my life to Jesus only days before, and felt a new lightness within, which was both exhilarating and energizing. I felt almost as though I was walking a foot above the hot, shimmering asphalt.

Entering the comparative darkness of the Masonic temple brought some relief from the heat. The large stone structure gave shelter from the sun. I was at the temple because I had been invited to a luncheon. This was not yet my home Lodge, as I had been a Freemason in neighboring Wisconsin and had just moved to Iowa a few months earlier.

Masonic jurisdictions are laid out in such a way that each state in the U.S. has its own Grand Lodge, and each one is autonomous. Although the Iowa Grand Lodge recognized my Grand Lodge in Wisconsin as legitimate, I would have to make some arrangements to join this Lodge in my new community. As yet, I was just a guest.

I'd attended one of the regular Lodge meetings in the evening, and had been challenged by the local

Lodge officers about my knowledge of the Masonic "ritual work" and my possession of a valid dues card. Both being in order, I was allowed to sit in on the ritual, and later invited to this luncheon, which was an opportunity for fellowship. I accepted with delight, feeling that it would be a good opportunity to make some acquaintances.

However, between the meeting and my arrival at the luncheon, I had made an extraordinary transition from one kingdom to another. God had moved upon my life in a miraculous way. Through a remarkable series of events I had knelt at my bedside, holding a crumpled Chick tract in my trembling fingers. That tract told me that all I had to do to be acceptable in Jesus' sight was ask Him to forgive my sins and be my Lord and Savior.

After a life of assorted metaphysical "highs" and "lows," I almost recoiled at this. Having spent nearly my entire life of thirty-four years jumping through religious hoops for what I thought was God, this seemed too quick and simple. Even as I knelt, I asked myself for the hundredth time: Can it actually be this easy? A whisper skimmed across my heart, sighing, Yes.

Finally deciding to take the Bible and God at their word, I surrendered myself to Jesus. I never knew how empty I was until Jesus filled me with His Holy Spirit! It was a new, "Born Again," infinitely improved model Bill Schnoebelen who entered this Masonic temple.

As I walked down the stairs of the temple to the dining room, I was filled with anticipation. I was excited to make some new friends in this city and was therefore not prepared for what was about to happen.

As I sat down at the long table richly laid out with china and food, I felt a curious dampening of the joyous quality which had so recently brightened my soul. I looked around the tables for some indication of what I was experiencing. Did anyone else sense this? About a hundred men sat around me, immersed in fraternal conviviality! Shaking hands and passing stories amidst the clatter of silverware and laughter, my "brothers" seemed quite unaffected.

As grace was offered, my spirit seemed to darken still further. At the conclusion of the invincibly non-sectarian prayer, we all responded in the customary Masonic fashion with a "So Mote it Be!" The words felt like wormwood on my tongue!

I do not believe I tasted anything that hour. My stomach felt as leaden as my soul. I could not quench the profound unease I felt. I had never experienced anything like this before at a fraternal gathering. The fellow to my right, a few years my senior, attempted to engage me in conversation. I tried to hold up my end as he told me about some special Shrine club to which he belonged after he learned I was a Shriner. However, my heart was not in the conversation.

By the time I listlessly began the dessert, there was little doubt about what was bothering me. Like what we Midwesterners call "heat lightning" at the horizon of my consciousness, the Holy Spirit was flashing the message to me through the haze of fraternal fellowship: FLEE THIS PLACE, MY SON!

I was bewildered by what I felt, and kept glancing around to see if any of the other men were exhibiting signs of disquiet. However, the joviality was in high gear! For the first time in nearly nine years as a Freemason, I felt like an invading microbe being attacked by some sort of antibodies! This feeling, disturbing though it was, had a deeper and more familiar apprehension—GUILT! For no apparent reason, I felt guilty about being where I was!

It was only very gradually that I was finally able to isolate this guilt. I felt the same way I had as a child when my mother caught me doing something naughty—gentle, patient, yet unmistakable reproof! FLEE THIS PLACE, MY SON!

Finally, I could abide it no longer. At the earliest moment, I made my excuses and extracted myself from the lunch. As I stepped outside once again into the clear sunlit afternoon, I felt inexplicably as if I was emerging from a cold, dank tomb. I crossed the street as quickly as I could and stopped by my car, trying to shake the clamminess which enshrouded me. I turned around and glanced back at the huge temple, feeling unexpectedly like Lot's wife in the Book of Genesis.

Perhaps it was my overwrought imagination, but as I looked at the white stone structure before me, shimmering in the cheerful sunlight, it seemed to settle into the earth slightly. I felt as if I was watching a lid being twisted into place on a specimen jar. I could almost hear the laughter and jokes of the men slowly turning thin and tinny as they realized the trap was tightening.

I felt as if I had barely escaped with my skin intact. Shuddering, in spite of the warm sun, I climbed into the car and thanked God audibly and earnestly for preserving me from whatever was going on. I still felt a spiritual "chill" when I arrived at my home.

This was a time of considerable spiritual searching for me. I had been saved by Jesus while a member of the Church of Jesus Christ of Latter-day Saints (LDS), known as the Mormons. I was still quite new in the Kingdom of God, and wondering if I should stay a Mormon. This led me to some intense Bible study made possible by the fact that I did not have a full-time job at the time.

I was desperately seeking God about whether or not I could remain in the Mormon church and still be faithful to this wonderful new relationship with Jesus Christ. It disturbed me deeply that I had not been nearly as spiritually troubled by attending the LDS church meetings as I had been by my visit to the local Masonic temple. I had not even considered Freemasonry in my religious "equation," since I had repeatedly been told by my Lodge brothers back in Wisconsin that Masonry was not a religion. Being a trusting soul, I had believed them.

However, my Bible studies and conversations with my Mormon leaders were convincing me that trust needed to be tempered with both Biblical knowledge and discernment. I sensed in my LDS church leaders an unease with my scriptural questions, and, if not quite a spirit of deception, at least a sense of them tap dancing delicately around issues rather than dealing with them. I was learning that my leaders were not being frank with me, and so my skepticism bled over into my attitude toward what I had been told about the Lodge.

I finally began finding things in the Bible which brought the sin in the Masonic Lodge into stark relief. The precious Holy Spirit was illuminating my mind as I prayed and fasted and sought His guidance. Scripture verses which I had read many times before suddenly flared up at me like fireworks in a darkened night.

I was beginning to see why I had felt so troubled in my spirit while seated at the Lodge luncheon. I kept finding clear Bible verses which indicted practice after practice within the Lodge rituals. As I became convicted of the sin of Freemasonry, I wondered what to do about it, beside the obvious decision to never go back to the Masonic temple.

Help came from a very unexpected source. As a Mormon, I had heard about a book which was a supposedly vicious attack on our church by a Mormon excommunicated from the church for adultery. I finally got up enough nerve to buy the book, which was called *The God Makers*. The book served to convince me that my misgivings about Mormonism were true, and that it was indeed an anti-Christian cult.

As time went on, I determined that the accusations made by church leaders about the authors, Ed Decker and Dave Hunt—especially Decker, the ex-Mormon of the pair, were outright lies to try and impeach his Christian testimony. That they would evidently spread rumors without batting an eye, did nothing to improve my opinion of the Mormon church leaders.

Surprisingly, the book also made several pieces fall together on the Lodge as well. In the chapters on the Mormon temple, the authors brilliantly drew parallels between the temple rites and Freemasonry. More importantly for me, it showed the occult and luciferian roots of Masonry in such a way that I knew even U.S. Masonry was steeped in hellbroth and damnation. I had thought that only the European varieties of Masonry were truly occult.

So *The God Makers* killed two birds with one stone, pretty much convincing me of both the Mormon cult's falsity and the Masonic Lodge's peril. The book provided me with a starting place to begin some serious research into dangers of the Lodge. My own background in occultism, witchcraft and even satanism prior to joining the Mormon church had provided me with a comprehensive knowledge of occult and esoteric (hidden, high-level) Masonry and ceremonial magick.

Additionally, I was deeply involved in American Masonry for over nine years and went through both Scottish and York Rite bodies, the Mystic Shrine and the Order of the Eastern Star. I held offices in a half-dozen Masonic bodies, including Junior Warden of my Blue Lodge, Kilbourne #3 of Milwaukee, Wisconsin and Associate Patron of the Eastern Star. Each of these offices demanded a high degree of zealous involvement, memory work and study. I had been a Masonic *fanatic*.

Now, by the help and grace of God, I was given the Biblical tools to evaluate all the experiences and knowledge that He had permitted me to amass during sixteen years in witchcraft and nine years in Masonry. This book is, prayerfully, the result of my

17

evaluation. It will look at the deep roots to which Freemasonry has sunk in the occult and witchcraft, and how it can truly be nothing more than a rival religion to Biblical Christianity, in spite of the protests of its leaders.

Without realizing it, Christian Masons have come to trust in the "lambskin" of their white leather aprons more than in the Lamb of God, slain from the foundation of the world (Revelation 13:8). They have lost their Biblical perception of the fact that, as they are worshipping at the altar of the Lodge before their "Worshipful Master," they are attempting to serve two masters: their true Master, Jesus Christ, and the Masonic "Master," who is but a fallible sinner like themselves. Jesus warned that:

> No servant can serve two masters: for either he will hate the one, and love the other; or else he will hold to the one, and despise the other. Luke 16:13

This is the sad dilemma that Christian Masons find themselves in, trying to continually choose their priorities. Almost invariably, Jesus is neglected—for the god of Freemasonry is a jealous taskmaster, while our Lord is gentle and patient with us.

But His patience is not without limit. It is apparent that we are reaching a crisis point in Western Christianity. As the world grows darker and more terrible with sin and despair, most denominations have forsaken the fountain of living waters and have hewn out cisterns of their own design— cisterns which soon become cracked and filled with the filth of the world's wisdom (Jeremiah 2:13).

Much of the heresy of Freemasonry has become the heresy of American Christianity. The feeble landmarks of Masonry will not be able to stand when the mighty winds of judgement begin to blow across the land. God will not wink forever at this strange fire being offered on His holy altar (Leviticus 10:1-3)! Those Christian men who have girded their loins in the lambskin of Masonry will suddenly find themselves startlingly naked before the judgement of the true God.

The cry of this book is from one who was in darkness, but has been brought to the Light of the world! It is a cry to every Mason who professes to be a follower of Jesus Christ. It is the same cry given by the apostle Paul many centuries ago:

> Be ye not unequally yoked together with unbelievers: for what fellowship hath righteousness with unrighteousness? and what communion hath light with darkness?
>
> And what concord hath Christ with Belial? or what part hath he that believeth with an infidel?
>
> And what agreement hath the temple of God with idols? for ye are the temple of the living God; as God hath said, I will dwell in them, and walk in them; and I will be their God, and they shall be my people.
>
> Wherefore come out from among them, and be ye separate, saith the Lord, and touch not the unclean thing; and I will receive you.
>
> And will be a Father unto you, and ye shall be my sons and daughters, saith the Lord Almighty. II Corinthians 6:14-18

1

The Best Thing That Ever Happened To Me!

I am often asked about the Lodge, and the kind of ministry I am involved in. People inquire:

> So what is the big issue with Masonry anyway? It's just a harmless bunch of guys running around in funny hats. Why are you so upset by them?

This question overlooks the terrible damage Freemasonry can and does inflict upon the Christian, and the church. It must be clarified, though, that Masons themselves are not the problem. They are but casualties in a larger battle. Masonry is an anti-Christian religion, and when Christians, especially Christian leaders join it, we should become alarmed.

We cannot expect unsaved men to know better. Their eyes are darkened by sin. They are what I was when I became a Mason—pagans. If they wish to join a pagan religion like Masonry, that is their affair.

However, it is an altogether different matter when Christians take part in pagan rituals. They should know better, but they either don't want to know the truth about their Lodge, or are too busy to look beneath the surface of Freemasonry to see if all the pious platitudes are true. God warns:

> My people are destroyed for lack of knowledge: because thou hast rejected knowledge, I will also reject thee, that thou shalt be no priest to me: seeing thou hast forgotten the law of thy God, I will also forget thy children.　　　　Hosea 4:6

That terrible warning has come back to haunt the families of countless Masons that we counsel in our ministry.

The father in the Christian home is supposed to be a priest to his family. But what happens to him and his family if he becomes steeped in the paganism of Freemasonry? A subtle spiritual breakdown commences. This is the most obvious reason why we warn those in the Lodge that all is not as it seems. A recent event illustrates my point.

I took a call at our ministry from a fellow who was quite upset. He had received some information on the dangers of Masonry, which we send out on request. He was beside himself and demanded that his name be taken off our mailing list. I told him we would, but I wanted to know what the problem was.

> "I'm a Mason," he proclaimed proudly, "and I don't appreciate getting this garbage that you sent! It's a bunch of lies!"

> "I see..." I managed to get in.

> "I don't mind telling you that I'm a Southern Baptist, and have been all my life, and the

22

Lodge has made a better Christian out of me than anything else I can think of!"

"Well, I'm sorry you're so upset, sir," I put in, "but I'd just like to tell you that I was a member of the Masons for nine years until I got saved, and then the Holy Spirit convicted me of the sin of it..."

"Listen, you!" he interrupted. "Freemasonry is the best thing that ever happened to me in my whole life, and..."

I jumped in. "Just a minute, sir! Masonry was the best thing that ever happened to you?"

"Yes!" he barked.

"Better than Jesus Christ?"

Silence.

"Are you telling me that Masonry has been a better experience for you than Jesus ?"

"Th...that isn't what I meant!" he finally managed to choke out.

"But listen to what you just said! What does that tell you about your spiritual life in...?"

"Just take my name off your mailing list!" he demanded, and hung up.

CONVICTION BRINGS ANGER!

That response is tragically typical. Christian Masons have been delicately, oh so gradually, blinded to their "first love," Jesus (Revelation 2:4). They would be the last to acknowledge it or admit it, but Freemasonry has taken the top shelf in their lives, and Jesus has been moved to second place.

Then, all of sudden, something prods their conscience, and it is like tearing the scab off an old wound. "Christian" Masons react with rage when someone says something against their Lodge—a rage often all out of proportion to what has actually been said. They feel the convicting power of the Holy Spirit at work in their lives, perhaps for the first time in years, and it chafes them. Their consciences, so long calloused and seared (I Timothy 4:2), have suddenly been rubbed raw by the truth.

The spiritual discernment, so long deadened, has been quickened to life. Like a foot which has "fallen asleep," their sense of right and wrong struggles gradually, painfully back to life, and it hurts! Anger is all too often the result and a retreat back into the comfortable lies of Lodgery. They listen to the siren song of sin whisper its lullaby in their ears: "All is well; nothing to fear."

Unless these Masons are prodded patiently, lovingly, and continually by the truth, they frequently never allow the painful truth to sink in. A spiritual numbness crawls over them like a warm blanket, and they roll over and go back to sleep.

A VITAL QUESTION

Why does this happen? Why do "godly" men, Christians who are often deacons, elders or even pastors, get so incredibly upset about the simple preaching of gospel truth? What is the big deal about Freemasonry? To answer these very important questions we will examine why Masonry is spiritually dangerous in the light of the only book that really counts, the Holy Bible.

What conclusions did I reach in my study of the Word of God which showed me that I could no longer be a Mason and still be a follower of my new-found Master and Lord, Jesus Christ? The main source of information is found in the Bible. But because of much confusion and misunderstanding about Freemasonry (some of it intentionally created), we will quote from the official books of Masonry when comparing their point of view to the Bible, as well as referring to recognized secular sources of authority.

What are these "official" books of Freemasonry? That's a touchy question because Masonry is secret. However, from my nine years in "the Craft" (as Masonry is frequently called), I believe the hierarchy of books used as references will cause every honest Mason to acknowledge it as pretty definitive:

1) **The Ritual Monitors.** These are the actual ritual workbooks, issued with the imprimatur of every Grand Lodge of each state. They are the highest recognized authorities, like the Bible is for Christians. We will quote from these ritual books, even though they are considered "secret" and protected by dire, bloody oaths, for we are commanded by the Lord to bring to light the hidden things of darkness (Ephesians 5:11).

These monitors are unique to every state, and there are slight variations among them. They are also unavailable to cowans (non-Masons), so we will use the DUNCAN'S RITUAL MONITOR, which is older, but is in print, and is substantially (95% or better) identical to the modern ritual work in Freemasonry.

2) **Authoritative writings.** If the monitors are the "bibles" of the Lodge, these authoritative writings would be the equivalent of the writings of

Church Fathers in Christianity. They are either references, philosophical books, or scholarly treatises
written by men of indisputable stature within
Masonry. Many of them are Thirty-third degree
Masons, such as Henry Wilson Coil, Albert Mackey
or Albert Pike.

Another book is the so-called "Masonic Bible,"
which is simply an edition of the Bible with the
Masonic seal stamped on the front and about 100
pages of additional illustrations and text showing
how Masonic teachings and legend can supposedly
be supported by the Bible.

Also included is the literature circulated by
official bodies such as pamphlets, etc., bearing the
official imprint of a Grand Lodge or the Supreme
Councils, or the York or Scottish Rites.

3) **Educational or philosophical writings**.
These books, the lowest in the hierarchy of definitive
authority, are the works of Masonic scholars or historians of lesser note—men who are Masons, and who
may even hold the coveted 33° rank, but who are not
the world-class experts as the authors above. These
books are designed to inspire, inform or edify the
Masonic reader.

A comparison of the Bible and Masonic books
will prove if Freemasonry is as harmless as it claims.

Jesus is God, and the commandments He gave us
in the New Testament are no less binding than those
in the Old. Let us not forget that all the Bible—every
single word—is God's inspired Word (II Timothy
3:16) and the writings of the apostles, like Paul, are
to be taken as seriously as those of Jesus Himself.

The Bible is our source book for what is, and is

not, sin. Even Masonry, in its rituals, teaches that the "Holy Bible is the rule and guide of our faith."[1] Therefore, we expect the Masons to agree with us that what the Bible says is sin, is sin indeed!

Sin is the doorway Satan uses to attack us and our families. If a Masonic father is doing something habitually sinful, without repenting, it may well open up the family to abuse from the realms of darkness! My opinions, or your opinions about sin don't count. It is the Lord's opinion that matters, and His opinions can be found most reliably in the Bible!

Hopefully, every Christian and even every Mason will give a hearty "Amen" on that one! With that premise in mind, and this groundwork laid, let us see how Freemasonry stacks up to God's standards.

PART 1

The "Religion" of Masonry

2

Can Masonry Really Be a Religion?

Over the years, Masonry has carefully tried to create a public image of being simultaneously not a religion and yet the "civil religion" of the United States.

Masonic institutions gain entry where no other religious denomination is allowed, doing things which, from any other source, would be considered a violation of the separation of church and state. Masons lay cornerstones of public buildings. They are allowed to give presentations in public schools of what they feel are wonderful virtues.

On the other hand, Masonry promotes itself as a religious organization, a society which makes "good men better." Masonry professes to be religious without being a religion. This assertion can be found in much of its public literature.

Freemasonry has very good reasons to present

itself as a benevolent, but non-religious society. Otherwise, it would alienate members of some of the most populous religions in the world, as well as the casual unchurched population who have no time for anything religious. In all fairness, we must ask if Masonry's claim is truthful. If Masonry is a religion, what does the Bible say about it?

WHAT IS RELIGION?

If the Lodge is just a "club," it would be harmless. Yet by all fair standards and by the words of its own scholars, it is a religion. Webster's dictionary defines "religion" thus:

> 1. ...belief in a divine or superhuman power or powers to be obeyed and worshiped as the creator(s) and ruler(s) of the universe.

> 2. ...expression of this belief in conduct or ritual.

> 3. ...any specific system of belief, worship, conduct, etc., often involving a code of ethics and a philosophy.

Does Masonry believe in a divine being? The answer is an emphatic *YES!* One of the standards for Masonic admission is the belief in a supreme being. This is borne out by the Ritual Monitor.

In the first degree (Entered Apprentice), the candidate is barely inside the Lodge door for his initiation when he is summarily challenged by the question: "In whom do you put your trust?" His answer must be, "In God."[1] Otherwise, he is not allowed to proceed with the initiation ritual. As a working Lodge officer, I was instructed of this, and was prepared to remove the candidate from the Lodge room if necessary.

Because Masonry requires the belief in a supreme being, it therefore has the first distinctive mark of a religion. The highly regarded Masonic authority, Henry Wilson Coil has written:

> Freemasonry certainly requires a belief in the existence of...a Supreme Being to whom he is responsible. What can a church add to that, except to bring into fellowship those who have like feelings? That is exactly what the Lodge does.[2]

IS THERE A MASONIC RITUAL?

Does Masonry express this belief in conduct or ritual? The answer to this question is *Yes*.. Masonry is highly ritualized, much more than most Christian churches. Lodge ceremonies are replete with prayers, funeral rites, and initiations.

A Lodge officer must memorize literally hours of ritual "work," or lines of recitation, which must be given letter-perfect at every Lodge meeting. So important is the precise wording, foot work, and rod-work that entire "Schools of Instruction" are provided, which officers are encouraged to attend. Masonic ritual experts from the Grand Lodge observe the ritual teams in action and critique them for minor infractions of either wording or gesture.

I attended such schools, and can vouch for the serious discipline. Masons conduct funerals, open and close every meeting with prayer, and celebrate initiatic rites of passage to higher degrees.

The Masonic author, Coil, makes an important point on this:

> Freemasonry has a religious service to commit the body of a deceased brother to the

dust whence it came and to speed the liberated spirit back to the Great Source of Light. Many Freemasons make this flight with no other guarantee of a safe landing than their belief in the religion of Freemasonry. If that is a false hope, the Fraternity should abandon funeral services and devote its attention to activities where it is sure of its ground and its authority. Perhaps the most we can say is that Freemasonry has not generally been regarded as a sect or denomination, though it may become so if its religious practices, creeds, tenets and dogma increase as much in the future as they have in the past."[3]

This certainly fulfills the second distinctive mark of a religion.

DO MASONS HAVE A CODE OF ETHICS?

Does Masonry have a system of belief, conduct or philosophy? Obviously it does, as the rites, the charities of the fraternity, and the volumes of writings make plain. Solemn norms of behavior are enjoined upon every Mason, especially one who has reached the Master Mason, or Third Degree. Masons are sworn to expect torturous death at the hands of their brother Masons as the immediate result of violating any of their rules.

Aside from belief in a god, the Masonic belief system also includes:

1) The immortality of the soul.

2) The judgement of Masons by the labors they have wrought.

3) A belief in extending charity and beneficence to all, and especially to brother

Masons and their families, widows and orphans.[4]

Do Masons have a code of conduct? The answer to that is affirmative. Above all else, Masons must keep the secrets of their brother Masons *inviolable*.[4] This discretion is commanded upon them in the most severe manner, and the oaths taken in the degrees are full of minute details about what may and may not be done by a Mason of a given degree.

For example, a Master Mason may not have "illicit carnal intercourse" with a brother Master Mason's wife, sister, mother or daughter. He may not cheat, wrong or defraud a Mason. He must never strike a Mason in anger, except in the lawful defense of family or property. His life must exemplify the Masonic virtues of "friendship, morality and brotherly love."[5]

This is quite evidently a highly developed, if somewhat selective code of ethics. It is obvious that Freemasonry bears all the distinctive features of a religion according to the dictionary!

How then do we respond to the Masonic PR that it's not a religion? Before we dismiss it as mere public relations fluff, let's see what is actually said "in house." Since the Lodge is a secret society, it might be informative to see what Masonic leaders write for Masons only, not for cowans and eavesdroppers[6] (non-Masons) to read.

If this sounds a bit paranoid, our ministry has several copies of the book, Morals and Dogma, by Albert Pike with the following printed on the frontispiece: ESOTERIC BOOK, FOR SCOTTISH RITE USE ONLY; TO BE RETURNED UPON WITHDRAWAL OR DEATH OF RECIPIENT.

OUT OF THE HORSE'S MOUTH?

Some of the most highly esteemed Masonic authorities shared their thoughts on the question of whether Masonry is a religion. Read the words of Albert Pike, 33°, called the "Plato of Freemasonry."[7] He was the former "Sovereign Grand Commander of the Supreme Council of Grand Sovereign Inspectors General of the Thirty-third Degree:"

> Every Masonic Lodge is a temple of religion and its teachings are instructions in religion.[8]

> It is the universal, eternal, immutable religion.[9]

> Masonry propagates no creed except its own most simple and sublime one; that universal religion taught by nature and reason.[10]

> That rite raises the corner of the veil...for there it declares that Masonry is a worship.[11]

Probably the leading historian and scholar of Freemasonry in modern times was Albert Mackey, 33°. He declared that "The religion of Freemasonry is not Christianity."[12]

These Masonic authorities, each regarded very highly by half the Grand Lodges in America,[13] agree that Masonry is a religion, and is not Christianity. The dictionary and the words of its own authorities and scholars declare it.

In addition to Mackey's observation that Masonry is not Christianity, we can do more than take his word for it. In the next few chapters, we will see if the religion of Freemasonry is somehow compatible with, or opposed to true Bible-based Christianity.

3

Is a Mason Serving Two Masters?

If Freemasonry is a religion, but not Christianity, then what is it? Can a Christian belong to two religions? Can he serve two Masters, the "Worshipful Master"[1] (title of the Lodge leader), and Jesus? Unless Masonry is identical to Christianity, it is a rival creed to Biblical faith!

Jesus says you cannot serve two masters! You will love the one and despise the other (Matthew 6:24). Jesus forbade His disciples from calling anyone "Master" (Matthew 23:8-10). The leader of the Lodge is not just called "Master," but "Worshipful Master!"

Jesus commanded, "Thou shalt worship the Lord thy God and him only shalt thou serve" (Matthew 4:10). The Bible (the rule and guide of every Mason) forbids the worship of any other god than Jehovah! Every book of the Bible enjoins the sole worship of the one God. Not only are Masons appropriating a title reserved for Jesus, they are giving their "Masters" an adjective that means they are worthy of worship!

"MIRROR, MIRROR ON THE WALL—WHO'S THE MOST WORSHIPFUL ONE OF ALL?"

The Mason must see how serious it is to worship anyone besides God. Remember what happened in the New Testament when someone tried to worship a man of God. In Acts 10:25, Cornelius was so overjoyed to have Peter come to his home that when the apostle walked in, he fell at his feet and worshiped him. But "Peter took him up, saying, Stand up; I myself also am a man" (Acts 10:26). Even when John, in the midst of extraordinary visions, fell at the feet of an angel, the angel gently warned him,

> See thou do it not: for I am thy fellow servant, and of thy brethren the prophets, and of them which keep the sayings of this book: worship God.   Revelation 22:9

Here is one of the greatest apostles in Christian history and an angel, both refusing to be worshiped even by men with the best of intentions. However, every Masonic Lodge has a "Worshipful Master."

It doesn't improve from there, either! Every Grand Lodge has a "Most Worshipful Grand Master" presiding over it. A Christian Mason should ask: If a Grand Master is most worshipful, then where does that leave God? As second-most-worshipful? A divine runner-up or "Mr. Congeniality?"

The Mason may protest that too much is being made of mere titles and claim that no one literally worships the Master of a Lodge. Should a Christian accept the kind of pharisaical hair-splitting that such a distinction requires? As a pastor advised me years ago, "Is it the business of a Christian to see how close he can camp to the 'borders of Egypt' (sin)? Or should he stay far away from those borders and remain well within "Canaan land?" Does not Paul give us the answer when he exhorts us to avoid "all appearance of evil" (I Thessalonians 5:22).

Jesus, Who is the Lord of every Christian Mason, evidently cared enough about "mere titles" to forbid their use in Matthew 23. If He forbids it, that should be good enough for any Christian. Yet, not only do Masons call their leaders "Master," they also refer to themselves as Master Masons!

It would seem that for the Christian in the Lodge, the dilemma boils down to either taking Jesus at His word, or choosing to ignore commandments of God in favor of the fraternity.

SERVING TWO MASTERS?

Christian Masons may protest that they don't actually "serve" the Worshipful Master of their Lodge. Although these men may function under that illusion, sadly it is only an illusion.

A Master Mason makes a solemn oath on the Bible that he will "acknowledge and obey all due signs and summons sent to me from a Master Masons' Lodge or given me by a brother of that Degree, if within the length of my cable-tow."[2] Does the Master Mason take this oath seriously?

If he does not, he is not much of a Mason. If he does, then he must take all of it seriously. He finishes the oath by saying, "All this I most solemnly and sincerely promise and swear...So help me

39

God..." He also swears that if he breaks any part of this oath, he will have, among other things, his body severed in two and disemboweled![3] Serious enough?

The Mason has sworn to obey all summonses sent from his Master. If that absolute obedience isn't being a "servant" of the Master, then the language has lost all meaning. If the Christian in the Lodge takes his oath in good faith, then he must indeed serve two masters: Jesus and the Lodge Master.

Ed Decker, author of *The Question of Freemasonry*, and I had occasion to interview two Freemasons at a state fair booth. Both were very pleasant, anxious to wax eloquent about the good done by the Lodge, and it was evident they were zealous for their Craft.

Towards the end of the conversation, we asked the men if they attended church regularly. Both hemmed and hawed a bit. One said he had been raised a Methodist but hadn't been to church in some time. The other said he hadn't had time for it lately.

It was evident that as the zeal of these men grew for Masonry, the fires of Christianity flickered and died, if they were ever lit.

As Ed and I walked away, we speculated about how unwilling those men would be to spend all day at the fair witnessing for Jesus. But there they were, promoting Freemasonry with a fervor that would shame any street preacher! Without realizing it, had they come to love the Lodge and despise the Lord Jesus Christ?

There is a deeper principle here—an uncomfortable Bible truth. Paul warns us:

> Know ye not, that to whom ye yield yourselves servants to obey, his servants ye

are to whom ye obey; whether of sin unto
death, or of obedience unto righteousness?
Romans 6:16

When we sin, we become servants of sin. The
more we sin, the more enslaved we become. This is
not news to anyone who has ever been entrapped in
sin. Paul says that "when ye were the servants of
sin, ye were free from righteousness" (Romans 6:20).

There are two possibilities here. All men are
slaves, and it just depends on which master we
choose to serve! We can labor in bondage to sin, or
we can become bond-slaves of the Lord Jesus Christ.
There aren't any other choices.

If Freemasonry draws you into breaking the
God's commands, it has made you a servant of
another master—a quite unpleasant one, whose
name is Lucifer.

Jesus beckons us, saying:

Come unto me, all ye that labor and are
heavy laden, and I will give you rest. Take
my yoke upon you, and learn of me; for I am
meek and lowly in heart: and ye shall find
rest unto your souls.　　　Matthew 11:28-29

That is a wonderful promise to the Mason,
especially since Masonic ritual puts so much
emphasis on labor and wages. The ritual teaches
that the reason one becomes a Master Mason is to
"work and receive Master's wages."[4] However, it is
never clear what those wages are.

In Romans 6, Paul offers an insight into the
nature of those wages:

What fruit had ye then in those things
whereof ye are now ashamed? for the end of

those things is death. But now being made
free from sin, and become servants to God,
ye have your fruit unto holiness, and the end
everlasting life. For the wages of sin is death;
but the gift of God is eternal life through
Jesus Christ our Lord.   Romans 6:21-23

This is one way Christianity can be distinguished
from other religions. All other Masters demand that
one work for wages towards rewards in the afterlife.
Jesus alone offers eternal life as a gift. Obviously, if a
Mason is earning something, it cannot be a gift.
Romans 4:4-5 says:

Now to him that worketh is the reward not
reckoned of grace, but of debt. But to him
that worketh not, but believeth on him that
justifieth the ungodly, his faith is counted for
righteousness.

The Christian Mason, who has already placed his
trust in the finished work of Christ, now also trusts
the wages of a Master Mason. But, what are those
wages?

Remember those "wages of sin" when we
contemplate the Master Mason oath which promises
disemboweling if the Mason fails to obey his Lodge.
Whether he knows it or not, the Mason is in bondage
to a religion whose wages are death. As the prophet
Ezekiel warns, "The soul that sinneth, it shall die."
(Ezekiel 18:4).

4

Masonry's "Generic" God?

You can tell a great deal about a religion by its god, and this is one of the central questions which must be analyzed in comparing Freemasonry with Christianity. Does the "God" of Freemasonry resemble the God of the Bible?

It is difficult to learn the name of the deity of Masonry, since it is a closely guarded secret! To outsiders, the god of the Lodge is usually referred to as "The Great Architect of the Universe" (or T.G.A.O.T.U.). That sounds alright, if a bit vague. This is, sadly, exactly how it is intended to sound.

The blandness of the titles of the Masonry's god diminishes as the degrees climb. As in most secret societies, the revealed material gets more bizarre as you get further into the club. As an example, in the first prayer the candidate hears, deity is addressed as "Almighty Father of the Universe."[1] Throughout the lower degrees, the deity is addressed either as "God" (as when the candidate swears his oath by saying, "so help me God...", or as T∴G∴A∴O∴T∴U∴.

As one progresses higher into the degrees, the nature of God begins to take on a less soothing quality. The closest description I can think of is generic foods in the supermarket. The god of Masonry is a "generic" god. His label is blank, so if you want to write in Allah or Krishna or even Satan, you could do so and no Mason could possibly object.

This is obviously "God-to-the-lowest-common-denominator." The Masonic authority, Albert Mackey put it this way:

> Be assured...that God is equally present with the pious Hindu in the temple, the Jew in the synagogue, the Mohammedan in the mosque, and the Christian in the church.[2]

One could just as logically add to Mackey's thought by saying that this "god" is equally present to the satanist in his coven as he cuts the heart out of a child. Lest the Mason think that such a statement is absurd, let us see just how discriminating the authorities are about the nature of the god they worship. Henry W. Coil, the most highly regarded Masonic scholar states:

> The Masonic test is a Supreme Being and any qualification added is an innovation... Monotheism has been espoused as the sole religious dogma of Freemasonry by some authors...This obviously violates Masonic principles, for it requires belief in a specific kind of supreme deity. [3]

Thus, if you tell the satanist that he cannot be a Mason because his supreme being, the devil, isn't up to snuff, you are in violation of "Masonic principles."

When I applied for Lodge membership, I was a witch, and attended an Episcopal church.[4] I was stupid enough to think the witch god, Lucifer, was the Supreme Being, the father of Jesus. So when two Masons came calling to check me out as a candidate, and asked me if I believed in God, I said yes, without hesitation, knowing that my god was Lucifer.

I was welcomed into the Lodge with open arms, and remained there for nine years. During that time none of my "Christian brothers" ever witnessed to me about Jesus. That would have been bad Masonic etiquette!

After a couple of years, I met two high ranking Masons who were also worshippers of Lucifer. One was a famous ritualist in the York Rite, and one was a Master of a Lodge.

A GOD COMMON ENOUGH FOR ANYONE?

This generic god is obviously a god with whom everyone can be comfortable, except Bible believers! He is a god who offends no one. However, is this broad-minded "man upstairs" the right God, the Biblical God?

The God of the Judeo-Christian heritage is not some blank slate upon which you can trace the form of whatever idol you like. He is very clearly described in the Bible. From Mt. Sinai, He thunders:

> I am the LORD thy God, which have brought thee out of the land of Egypt, out of the house of bondage. Thou shalt have no other gods before me;   Exodus 20:2-3

45

Later in the Ten Commandments, He warns:

> Thou shalt not bow down thyself to them
> (other gods), nor serve them: for I the LORD
> thy God am a jealous God, visiting the
> iniquity of the fathers upon the children unto
> the third and fourth generation of them that
> hate me;   Exodus 20:5

Imagine the colorless god of the Lodge being jealous? It just doesn't wash. The God we quoted is not pleased to be identified with another god. He insists throughout the Bible that He is the only true God and He will tolerate the worship of no other deity (Deuteronomy 6:4, Isaiah 43:10, 44:6-8, I Timothy 2:5).

A "MR. POTATO-HEAD" GOD?

What do the Masons think of such a god?   Coil writes:

> Men have to decide whether they want a god
> like the ancient Hebrew Jahweh, a partisan,
> tribal god, with whom they can talk and
> argue and from whom they can hide, if
> necessary, or a boundless, eternal, universal
> undenominational and international, divine
> spirit, so vastly removed from the speck
> called man that he cannot be known...So
> soon as man begins to laud his god and
> endow him with the most perfect human
> attributes such as justice, mercy, beneficence,
> etc., the divine essence is depreciated and
> despoiled.[5]

Can we stop here a moment and unpack this a bit? Like much of the writing of Masons, it takes a bit of analysis to cut through the fog factor.

First of all, we are told that men may decide

46

what sort of a god they want. One has an image of the Mason passing by some celestial smorgasbord. Let's see, we like the mercy of Jehovah, so let's take it. But we like the ethics of Shiva over there, so let's choose those. We need luck, so let's choose Ganesh back here.

This approach to God is blasphemous. Theologically, you end up with a crazy quilt deity—the eyes from one god, the nose from another, the ears from a third, if you will. You end up with a kind of "Mr. Potato-Head" God. If you tire of his ears, you just pull them out and stick in a different set.

This sloppy concept of God is not unique to Masons. Sadly, it has become the prevailing way of thinking of God in secular culture, and the Masonic influence in the U.S. on "civil religion" may have something to do with it!

A MASONIC CONSENSUS

Coil is not alone in delineating this Masonic theology of God. Albert Pike, another highly esteemed authority proclaimed that:

> Masonry, around whose altars the Christian, the Hebrew, the Moslem, the Brahmin, the followers of Confucius and Zoroaster, can assemble as brethren and unite in prayer to the one God who is above all the Baalim, must needs leave it to each of the initiates to look for the foundation of his faith and hope to the written scriptures of his own religion.[6]

Note that here, as in Coil, Masonry's God is a deity all religions can worship. This is fascinating in itself, because the gods of most of these religions have had their followers killing each other for centuries. What possible deity can all these diverse types of people worship in common, except a god so faceless and void as to be utterly without meaning?

In the preceding quote, the God of the Bible is lumped with all these other deities and referred to as part of the "Baalim." Pike then says that the god of Masonry towers over them all. The term, Baalim, means false god. So Pike is calling the God of the Bible a false god, one who is subservient to the god of Freemasonry. Elsewhere, Pike generously wrote:

> (Masonry)...reverences all the great reformers. It sees in Moses, the lawgiver of the Jews, in Confucius and Zoroaster, in Jesus of Nazareth, and in the Arabian iconoclast, great teachers of morality and eminent reformers, if no more; and allows every brother of the order to assign to each such higher and even divine characteristics as his creed and truth require.[7]

Big of him! Especially since all the men mentioned in his broad-minded statement have been dead and rotting in their graves for centuries, except for Jesus of Nazareth! But the Mason can never confess that glorious truth!

Another modern Masonic writer, Manly P. Hall, 33°, agrees with the above writers:

> The true Mason is not creed-bound. He realizes with the divine illumination of his lodge that as a Mason his religion must be universal: Christ, Buddha or Mohammed, the name means little, for he recognizes only the light and not the bearer. He worships at every shrine, bows before every altar, whether in temple, mosque or cathedral, realizing with his truer understanding the oneness of all spiritual truth.[8]

Notice that the Mason is not "creed-bound." This means that all Masons who stand in their churches and profess the Apostles Creed, or have signed doctrinal statements to be members of a church, are either not very good Masons, or they are falsely representing their Christianity. Are not true Christians expected to be "creed-bound?"

Also consider the fact that Hall, like Pike before him, shows that somehow the Mason has a higher understanding than those who worship at Christian, Buddhist or Islamic altars. Thus, whatever the Christian Mason may believe outside of his Lodge, his beliefs within the Lodge are more lofty and true.

PICKING OUT A NICE GOD?

Coil's approach turns the Biblical concept of God backwards! We don't pick and choose a god like buying a new car! Jesus tells us in John 15:

> Ye have not chosen me, but I have chosen you, and ordained you...If ye were of the world, the world would love his own: but because ye are not of the world, but I have chosen you out of the world...   John 15:16,19

Most of us are familiar with Matthew 20:16 where the Lord says, "many are called, but few are chosen." We don't choose God. He chooses us! If we are willing to settle for the patchwork god Coil describes, we have substituted a creature of our own minds for the Almighty Creator (Romans 1:22-23).

Notice also that God is singled out by name as "Jahweh." He is referred to by uncomplimentary terms like a "tribal god." This makes Him sound like some little idol sitting on a rock somewhere in heathendom being worshiped by ignorant louts.

He is described as being a God with whom one can converse, or even argue. This much is true. But He is also described as a God one can hide from. Obviously, Coil never read much of the Bible which lies in such glory upon his Masonic altar. If he did, he might have found Psalm 139:7-10 which says of the Biblical God:

> Whither shall I go from thy spirit? or whither shall I flee from thy presence? If I ascend up into heaven, thou art there: if I make my bed in hell, behold thou are there. If I take the wings of the morning, and dwell in the uttermost parts of the sea; even there shall thy hand lead me, and thy right hand shall hold me.

David goes on to proclaim that God even saw him in his mother's womb (Psalm 139:13-14). Surely, we cannot hide from this God.

AN IDOLATER'S CONJURING TRICK?

Having denigrated God, Coil then describes his deity, a "boundless, eternal, universal, undenominational...spirit"—so far removed from men that he cannot be approached. He also has no qualities of personality. He is neither just, nor merciful. He may not even be a "he!" He sounds like an "it."

Coil says that to ascribe qualities like love to deity is to depreciate or despoil it. In that case, Jesus must have been one whale of a despoiler! He taught that God loves us, that He is merciful, just and compassionate. He is also a God who gets angry at evil. The Bible testifies of a God who gets really worked up about humanity, a God who grieves, who laughs, who rejoices—in short, a passionate God. Most of the writings of the prophets show a God who is anguished over the sins of His people, very much like a father who stays up all night consumed with grief over a wayward child.

The prophetic and apocalyptic books talk about the wrath of God. The image is one of a seething pot ready to boil over. This is not the placid, unapproachable deity which Coil extols. This is a vital, caring Being who is intimately engaged in the affairs of His sinful but beloved creatures, Who is not afraid to (pardon the expression) "spill his guts" in passionate expressions of concern.

Interestingly enough, this dispassionate, long distance deity of Coil's is very similar to the God-force represented in the New Age movement. It is no coincidence that until 1990, the name of the official Scottish Rite Masonic journal was *The New Age.*

This god is not that very different from the deity extolled in Islam as Allah, or in witchcraft. Part of the reason for the resemblance is that this nebulous "it" is so vague that almost any deity could fit, except the God of Bible.

What Coil has done is employ the old idolater's conjuring trick. First, he proclaimed a god so unknowable it becomes totally irrelevant to human beings. Then, he turns around and says, in effect:

> "Since this god is so far away and hard to understand, let me provide you with one a little closer to home which represents the faraway, inscrutable deity. Let this (fill in the blank here: 1. tree, 2. rock, 3. idol, 4. letter "G") stand as a symbol of the unknown god, and we can worship it."

This trick is as old as sin, but it is one Satan never

tires of foisting upon the children of men. Create a god so vague that it is incomprehensible, then create an intermediary idol to represent the deity and receive its worship. Whether it is the Mason kneeling before the square and compass on his altar, the pious Hindu praying to an idol of Brahman or the devout Catholic praying to St. Jude for a lost keychain, the end result is the same. They have:

> ...changed the glory of the uncorruptible God into an image made like to corruptible man, and to birds, and fourfooted beasts, and creeping things. Wherefore God also gave them up to uncleanness through the lusts of their own hearts, to dishonour their own bodies between themselves. Romans 1:23-24

Although the Masonic writers make their theology sound very philosophical, it is simply Satan's old shell game of moving God far out of sight, denying Him any practical value, and then substituting an idol in His stead.

There is no denying that the Biblical God is far beyond human understanding, and His ways far above our own. The difference between Him and this generic deity of the Lodge is that He has taken a very important initiative in humanity's direction.

He bridged the gap for us. He sent His Son (very God of very God and yet fully human), into the world to touch us, to suffer among us and to save us. Jesus Christ is like us in all ways, except sin. (Hebrews 4:15) He cared enough to come into our world and take our suffering upon His shoulders.

No other god makes that claim, and no other god would dare to!

5

The True Name
of the God of Masonry

In advertising, there is a practice known as "bait and switch." A store runs an ad for an item at a fantastic price. However, when the consumer gets to the store, he discovers the item is sold out. But, the store just "happens" to have another item in stock which is very similar, but is not on sale.

Human nature being what it is, the shopper is so geared up to buy the item he came for that he will buy the more expensive one rather than do without.

In Freemasonry, this practice is carried on all the time. The only difference is that in dealing with religion, the Better Business Bureau cannot get involved. What a pity!

The candidate, from the first degree on, is led to believe he will have imparted to him important secrets—secrets so earth shattering they must be protected by solemn oaths on the Bible, and serious enough to be covered by death penalties. After swearing the oath, he learns a secret handshake and a word which is easily found in the Old Testament (Boaz).[1] A similar set of "secrets" is imparted to him in the second degree.

Both degrees cost money (not peanuts, either). When I was going through them in the mid-70's, they cost around $50 apiece for the Blue Lodge degrees. That means the second degree Mason is out at least $100!

The candidate's anticipation builds. In second degree he learns what the liberal arts are, and that the letter "G" stands for God and geometry (or he thinks he learns these things). The liberal arts can be found in any encyclopedia, and it does not take a genius to figure out that geometry and "God" are spelled with a "G."

When the candidate hands over another fifty dollars or so, and learns an arduous amount of memory work, including the blood oath of 2nd degree which is six long paragraphs in length, he is ready for the ultimate secret of Masonry (or so he thinks). Here is where the "bait and switch" comes along.

He is taken through a bone-rattling 3rd degree initiation. Then after all this, is told the great SECRET of Masonry, the sacred name of God, (which is guarded by hours of ritual and three chilling oaths) has been forever lost! He has just shelled out at least $150 and invested months to learn that the "Master's Word" has been lost.[2] Instead, he is given a substitute word, "Mah Hah Bone", which is whispered into his ear while embracing the Master of the Lodge on the Five Points of Fellowship.

LEARNING THE TRUE NAME OF GOD

Most Masons don't really know what to expect in the way of secrets. Perhaps most aren't really disappointed, since most of them joined the Lodge for more pedestrian reasons: to go on and join the Shrine (where all the parties are), or to further a professional career.

However, if you hang around the Blue Lodge meetings long enough (usually a couple of weeks will do), you learn there are higher degrees that the Master Mason can take. These degrees proceed in one of two fashions in American Freemasonry. Usually, one of your brother Masons will encourage you to join either the York Rite or the Scottish Rite. Here, you are told, you will learn the really valuable secrets.

This is the "bait and switch," since these higher degrees invariably cost more money. The cost per degree is less, but both forks in the road of Freemasonry can run the Mason into several hundred dollars over and above what he has already paid.

If the new Master Mason is perceived to be a Christian, quite often he will be directed to the York Rite, since that has the "Christian degrees." If the Mason is more secular, or perhaps in a bit of a hurry, he is advised to go the route of Scottish Rite, which rockets you through 29 degrees in a couple of weekends, and enables you to go on and join the Shrine.

The Shrine is even more expensive, but it is the "fun" part of Masonry, we are told. Men are allowed to drink booze there, and ride around in funny little cars in parades—hopefully not at the same time!

In these two "higher bodies," the content of Masonic secrets begins to take on a more solemn tone. In the York Rite, especially, the Masonic candidate realizes he is going to acquire knowledge of a profoundly mystical nature. He will learn the true name of God!

This is supposedly the "Master's Word" which was forever lost, but is miraculously recovered four degrees (and a couple of hundred dollars) later. That could excite almost anyone, even the Mason who is in it only for the parties, or the influence.

In the Royal Arch degree, the centerpiece of the York Rite, the candidate is taken through a drama in which he supposedly enters a chamber beneath the ruins of King Solomon's temple during the time of the return of the Israelites from the Babylonian exile.

Within this chamber, he and two companions discover the lost Ark of the Covenant. On top of the Ark is a golden plate upon which is engraved the "Grand Omnific Royal Arch Word."[3] This is written in an arcane cypher alphabet which the candidate cannot read. He is told it is the name of God in three languages.

Under the "Royal Arch," a special position involving three "companions" of the Royal Arch, this word is communicated as the ineffable name of God—lost with the death of the Grand Master Hiram Abiff in the Third degree, but now restored in the Royal Arch degree.

The name given is JAH-BUL-ON. The High Priest of the Royal Arch says that this is "the divine

Logos, or "Word" referred to in John 1:1-5."[3] This odd name is supposedly the true name of the deity of Freemasonry, revealed at last! It is so "sacred" that it can only be revealed in the presence of three Royal Arch Masons while kneeling under a "Royal Arch" formed by their intertwined hands! This is pretty heavy stuff.

THE UNHOLY TRINITY

The Royal Arch candidate is led to believe this is the lost name of God which used to be pronounced with great solemnity by the high priest of the temple in the Holy of Holies. This is reflected in the fact that the senior office in a Royal Arch chapter is called High Priest, and even wears a mock-up of the Jewish high priest's regalia when ritual work is done.

It is the ineffable name of God (at least the Masonic god) and is quite an awesome secret indeed. However, in our search for the god of Masonry, it is also a major clue. The analysis of the name as presented to the candidate reveals some disturbing associations for anyone concerned with Biblical truth and proper reverence for the name of God.

JAH (the first syllable) represents the name Yahweh or Jehovah, the name of the God of Abraham, Isaac and Jacob. This name is so highly revered that it appears only a few times in the Authorized Version (1611), as in Psalm 68:4 and in the use of the word, Hallelujah, which means "Praise Jah!" So far, so good.

BUL (the second syllable) represents the name Ba'al or Bel. This is the name of the god of Jezebel and Ahab, perhaps the most wicked couple ever to sit on the throne of Israel (I Kings 16:29-33).

ON (the third syllable) represents the name of the Egyptian sun god. It is the name of his sacred city, Heliopolis, (city of the sun in Greek) in Egypt (Genesis 41:45, 50).[4] That is the god of Pharaoh!

🏹 I wonder how happy the God of the Bible is with the blasphemy of His name—a name so holy that people were stoned to death for taking it in vain—so holy that it could only be spoken in the Most Holy Place of the temple—so holy that pious Hebrew scribes down through the centuries dared not write it in their texts out of wondrous awe, and instead, substituted the name "Adonai" or "Lord." This is the same name of which God said:

> Thou shalt not take the name of the LORD
> thy God in vain; for the LORD will not hold
> him guiltless that taketh his name in vain.
>
> Exodus 20:7

How pleased do you think God is with this same name thrown into a metaphysical trash compactor and smashed together into a mystifying muddle with two of the most notorious idols of the Old Testament?

It should not require a Bible scholar to see that the god represented by this "Grand Omnific Royal Arch Word" is a very unholy trinity. It also represents a perfect distillation of the attitude towards God, as noted in the last chapter where it was made clear that just about any old god would do for a Mason. Here that concept is given concrete reality.

Remember that the ritual above identifies this Jah-Bul-On with the "Word" of John 1. Even the most cursory reading of John 1, especially verse 14, reveals that the Being discussed is Jesus Christ, not this three-headed theological monstrosity! If anything, it is more blasphemous than ever to identify Jesus, whom Freemasonry refuses to worship, with a deity that is two-thirds satanic!

Obviously, this Jah-Bul-On fellow is not the God of the Bible. And if he (or it) is not God, he must be a false god—a mask for Satan.

THE ANGEL OF THE BOTTOMLESS PIT

This is a pretty provocative statement, and sadly, it is not hard to find collaboration for it. For a start, we will briefly examine the other "fork" on the Masonic road, the Scottish Rite.

In the degree of the Knights of East and West (17°) in the Scottish Rite, there are many unwholesome and evil teachings, but we will cover only one. As Jah-Bul-On is the sacred name of the Royal Arch degree (York Rite), so there is a "Sacred Word" in the 17° as well.

This name is Abaddon.[5] A quick trip to the New Testament will reveal that, according to God's Word, there is nothing sacred about the name Abaddon.

> And they had a king over them, which is the angel of the bottomless pit, whose name in the Hebrew tongue is Abaddon, but in the Greek tongue hath his name Apollyon.
> Revelation 9:11

Both these terms mean "destroyer" and the angel of the bottomless pit is Abaddon, another word for Satan, whom Jesus identifies (John 10:10) as one who comes to steal, kill and destroy! Some sacred word!

This is not the only evidence for the identity of Masonry's god. Several ambiguous (and not so ambiguous) references are made in the writings of some of the Masons quoted before. Albert Pike, the highest Mason in the U.S in his day, wrote this:

> Lucifer, the Light-bearer! Strange and mysterious name to give to the Spirit of Darkness! Lucifer, the Son of the Morning! Is it he who bears the Light, and with its splendors intolerable blinds feeble, sensual or selfish Souls? Doubt it not! for traditions are full of Divine Revelations and Inspirations: and Inspiration is not of one Age nor one Creed. [6]

Although that statement is a marvel of ambiguity, we must remember that the thing sought in every Masonic initiation is "LIGHT." The question asked of the initiate at the critical moment is what he most desires. Prompted by the Deacon, the candidate says "Light" in first degree, "more light" in second degree and so on.[7] It is not without significance that this eminent Mason tells us that Lucifer is the source of that light.

Elsewhere, writing of Satan, Pike teaches that he:

> ...is not a person, but a Force, created for good, but which may serve for evil. It is the instrument of Liberty or Free Will. They represent this Force, which presides over the physical generation (i.e., sex), under the mythologic and horned form of the God PAN; thence came the he-goat of the Sabbat (witches' festival), brother of the Ancient Serpent and the Light Bearer or Phosphor of which the poets have made the false Lucifer of the legend.[8]

Pike is also quoted as giving this instruction to a council of very high level Freemasons:

> The Masonic Religion should be, by all of us initiates of the high degree, maintained in the

purity of the Luciferian doctrine. If Lucifer were not God, would Adonay (sic) whose deeds prove his cruelty, perfidy, and hatred of man, barbarism and repulsion for science, would Adonay and his priests calumniate him?

Yes, Lucifer is God, and unfortunately Adonay is also god. For the eternal law is that there is no light without shade, no beauty without ugliness, no white without black, for the absolute can only exist as two gods: darkness being necessary for light to serve as its foil as the pedestal is necessary to the state...

Thus, the doctrine of Satanism is a heresy; and the true and pure philosophical religion is the belief in Lucifer, the equal of Adonay; but Lucifer, God of Light and God of Good, is struggling for humanity against Adonay, the God of Darkness and Evil.[9]

Pike is not alone in beating the drum for Lucifer. The Masonic scholar, Manly P. Hall writes:

When the Mason learns that the Key to the warrior on the block is the proper application of the dynamo of living power, he has learned the Mystery of his Craft. The seething energies of LUCIFER are in his hands and before he may step onward and upward, he must prove his ability to properly apply this energy.[10]

SLIPPING OFF THE MASKS

Masons are on a quest for light. But it seems they have done everything they can to detour around Jesus who said "I am the Light of the world." These quotations may stun the person who thinks of

Masons as a nice order of men who help crippled children.

It has been amply demonstrated from authoritative sources, the Ritual Monitors, that Freemasonry does not worship the God of the Bible.

Frankly, that only leaves one other being they can worship—whether they call him the Great Architect or Jah-Bul-On. It is still Satan residing behind the masks of these other names.

6

Who Is Jesus to the Lodge?

This is a question that has haunted the ages. As Christians, we believe the Bible has the authoritative answer. Contrary to a century or so of liberal "Christianity," secularism and New Age twaddle, the Bible says Jesus Christ is unique. He is Almighty God come in the flesh! (John 1:1,14, Colossians 1:15, 2:9, etc.)

With the re-emergence of Masonry as a force to be reckoned with in our churches, we have seen a proliferation of different versions of Jesus.

We have the "Jesus" of the film *The Last Temptation of Christ*, a deluded, weak-kneed lecher, and the rock "Jesus" of *Jesus Christ, Superstar*, who dies and never rises again. There are the more dangerous "Jesuses" of the Mormon church and the Jehovah's Witnesses. Then there is the New Age "Jesus," an ascended Master who had to reincarnate many times and who needs to report back to Maitreya.

Suddenly, there are more Christs out there than

you can shake a stick at! It seems individual Masons and church denominations which have opened the door to Freemasonry have simultaneously shut the door in the face of the Holy Spirit.

They are trying to eat at the table of the Lord and the table of devils at the same time (I Corinthians 10:21). It doesn't work! We must find what the Masonic answer is to Jesus' all-important question from Matthew 16:15: "Whom say ye that I am?"

THE CASE OF THE MISSING CHRIST

There is no mention of Jesus in the Blue Lodge ritual. Although prayers are offered at each Lodge meeting and in each initiation rite, Jesus may not be mentioned!

Instead, the Masonic rites climb to a crescendo of praise to a fellow described as an example of an "instance of virtue, fortitude, and integrity seldom equalled or ever excelled in the history of man." This man is said to be "one of the greatest men, and perhaps the greatest Mason, the world ever knew..."[1] Instead of Jesus, the central figure of the Lodge is a fellow called Hiram Abiff.

If you know your Bible well, you may recognize this Hiram as the temple artisan mentioned in I Kings 7 and II Chronicles 4. The REAL Hiram is a tiny thread in the vast warp and woof of the fabric of God's Word.

For the Mason, Jesus is shunted aside in favor of Hiram, the widow's son. A Mason is not allowed to pray to or testify of Jesus in the Lodge. A Christian Mason cannot even share the joy of Jesus with a "brother Mason" in the Lodge. We need to examine why Jesus is so strangely absent from Masonic ritual.

64

If Masonry is a religion, and it certainly is, then we must ultimately measure it by what it thinks of Jesus Christ. If the Lodge's view of Jesus conforms to the Bible, there is no problem. But to the extent that the Masonic Jesus moves towards all these other strange Christs, to that extent we have heresy!

A GHETTO FOR JESUS?

I can already hear the howls of protest from Masons, saying Masonry is a Christian institution, and that there are Masonic degrees that require one to be a Christian to qualify.

I received a letter recently from one of hundreds of Godly wives who write our ministry seeking help in prying their husbands loose from the Lodge. She had shared our tracts with him, but he said "you couldn't be a Mason without being Christian." He also told her Jesus "was the the Great Architect." Unfortunately, both statements were incorrect!

When I read letters like that, I weep. I don't know if this poor woman's husband is too spiritually dull to understand the mess he has stepped into by joining the Lodge, or if he is following Masonic procedure and lying his lips off to her!

I pointed out to this woman that there is a sizeable difference between Jesus, as the Creator of the universe (John 1:1-3) and Jesus as a cosmic carpenter. To create means to make something out of absolutely nothing. But a builder or architect must work with raw materials, lumber and cement. To call Jesus the Great Architect is to belittle His awesome power to call a universe into existence out of nothingness!

There is a part of Freemasonry which requires a

person to be a nominal Christian to join. This is the Commandery, the Knights Templar (KT). It is the pinnacle of the York Rite.

I call this degree the "Jesus ghetto." Just as the unfortunate Jews in Europe were forced to live in cramped ghettos by persecution, and today many live in ghettos because of poverty, so the Lord Jesus is squeezed into one, measly degree out of a panoply of some forty-three degrees!

It is the one degree where prayers are offered in Jesus' name, and where the cross is part of the symbolism. Remember, though, this is the same York Rite that worships Jah-Bul-On, and Chapter officers blasphemously apply the holy name of God, "I AM THAT I AM" to themselves at every meeting.[2]

This is the same York Rite that insists its companions take blood oaths committing themselves to have their ear sliced off, their tongue split from tip to root, their heart ripped out and placed on a dunghill to rot, or their skull smote off and their brains exposed to the rays of the noon day sun, should they ever violate their oaths![3] Nice Christian values!

While there is no doubt that external Christian piety oozes from the KT ritual, right down to singing *The Old Rugged Cross*, it needs to be emphasized that externals do not make an institution Christian.

I am a prime example. I was admitted to the Commandery, though I didn't know Jesus from a doorknob! Like so many, I thought I was a Christian because I had been sprinkled with water as an infant. Nothing in the Commandery ritual showed me the error of my ways!

What kind of Christian institution allows a person who is not a Christian to sit in its midst for years, and even assume the office of Prelate (a Chaplain) and never bother to find out if the person is really a believer.

TRODDING UNDERFOOT THE SON OF GOD?

The sad fact is that the resemblances between the Commandery and Christianity are just cosmetic. First of all, the extremely martial character of the order is a bit out of place for an institution supposedly following the Prince of Peace. Much of the order's ritual work consists of quasi-military drills and waving swords around.

Additionally, the KT ritual clearly avoids the Christian gospel of grace (Ephesians 2:8-9). Instead, the candidate for initiation is taken through a supposed seven-year "pilgrimage" as a "Pilgrim Penitent"[4] and a "Pilgrim Warrior."[5]

Rather than being saved by the blood of Jesus (which is barely mentioned), the candidate is taught he is saved through works of penitence (a word not even found in the Bible) and through making lengthy pilgrimages and fighting to defend the "Christian religion."[6]

A SINISTER SACRAMENT?

The high point of the KT initiation is when the candidate is brought before a large, triangular table covered in black velvet illuminated by candles and containing eleven silver goblets and a human skull enthroned on the Bible.[7] (Skulls figure prominently throughout the initiation.)[8]

This is intended to be the Last Supper.[9] It seems

but a grim mockery, though. The visual effect is more satanic than Christian, especially to one accustomed to the Table of the Lord in churches. However, the ambiance is the least of the problems.

The candidate is asked to partake of five libations, (toasts). The first three libations are given, respectively, to the memory of Masonic heroes King Solomon, Hiram, King of Tyre, and Hiram Abif.[10] The fourth libation is to the memory of Simon of Cyrene, and the fifth is the most sinister of all.

The candidate is never told to whom the fifth libation is drunk (it is "sealed"), and it is offered to him out of a human skull! He is told by the "Eminent Commander" to repeat a short oath which says, in part:

> as the sins of the whole world were once visited upon the head of our Saviour, so may all the sins of the person whose skull this once was, in addition to my own, be heaped upon my head, and may this libation appear in judgement against me, should I ever knowingly or willingly violate my most solemn vow of a Knight Templar, so help me God...[11]

Can you imagine any Christian taking that oath, thinking he was partaking in a "Christian order?"

Consider that the enormity of ONE SIN is enough to send a human being to eternal hell! You then solemnly swear in God's name that all the sins of another person, PLUS the sins of your own life which were (if you are a Christian) bought and paid for by the blood of Jesus, should be re-applied to your life!

The Holy Spirit has something to say about people who do such things:

> He that despised Moses' law died without mercy under two or three witnesses: Of how much sorer punishment, suppose ye, shall he be thought worthy, who hath trodden under foot the Son of God, and hath counted the blood of the covenant, wherewith he was sanctified, an unholy thing, and hath done despite unto the Spirit of grace?
>
> Hebrews 10:28-29

To "sell back" to Satan, under solemn oath, sins that Jesus died for already, is an incredibly heinous slap in the face to the very Lord the Templars profess to venerate!

Though most who take this oath do not understand what they have said, and think it is a light and frivolous thing to do, its spiritual consequences are devastating! This is a blasphemy on the Lord's Supper—an unholy parody almost as bad as the Satanic Mass. It is undoing the very covenant that Jesus began (Matthew 26:28).

A Christian KT who partakes of this evil communion is drinking "the cup of the Lord, and the cup of devils" (I Corinthians 10:21). Paul warns us that:

> Wherefore whosoever shall eat this bread, and drink this cup of the Lord, unworthily, shall be guilty of the body and blood of the Lord.
>
> I Corinthians 11:27

All this is on top of swearing an earlier oath that his head might be "smote off and placed on the highest spire of Christendom."[12]

This is not Christianity, but a carefully devised fraud; and a fraud which one can join only after taking 12 Christ-denying oaths! Yet many Christian leaders are proud members of the Commandery!

MICKEY MOUSE MONEY?

An effective counterfeit bill must look as similar to real money as possible. Only a fool would spend all his time counterfeiting bills and then put Mickey Mouse's picture on them.

Yet people forget this when they move into the spiritual arena. For some reason, they assume that because something resembles Christianity, it is genuine Biblical faith! Nothing could be further from the truth. Whatever else Satan may be, he is no fool! He is not going to create the religious version of a Mickey Mouse dollar bill!

Like the counterfeit twenties that our Treasury department must deal with, Satan's spiritual "funny money" is 98% perfect! A counterfeiter will try and create the perfect plates, find the exact same colored ink and the exact kind of paper with the same level of rag content. Why don't people assume Satan will do the same thing?

Some counterfeit bills can only be detected by Treasury department experts! To use a popular analogy, those agents know a fake bill because they have studied real money until their eyeballs pop! Similarly, true Bible believing Christians must study the Bible thoroughly and daily to detect Satan's counterfeit.

As believers, we must give Satan credit for being at least as subtle as some crook! He is not going to open a Masonic Lodge and hang a sign on the door saying, "First Lodge of Satan: Abandon Hope All Ye Who Enter Here." He would not get many members. No, he would strive to his uttermost to make certain that his Lodge looks good. Yet, because

70

he is the devil, he will always make mistakes and leave tiny clues, chinks in his deception, so that a person who knows the Bible will detect them.

Is God that picky? It is more a matter of Satan being the ultimate legalist, and using any tiny hook he can to get into you. Spiritual "tolerance" factors can be very high.

A friend of mine was taken on a tour of a huge hydroelectric plant in our state. He was shown turbines weighing hundreds of tons. The guide explained that they didn't dare touch their naked fingers to the area where the parts of the core meet the outer hull of the machine. The oil from a single fingerprint could unbalance and destroy the hundred-ton turbine. The machined tolerances were that precise!

God is a much better "engineer" than any human being, and we are the jewel of His creation. The Lord "machined" us with a precision that makes a Swiss watchmaker look like a fumble-fingered clod! As the psalmist said, we are "fearfully and wonderfully made" (Psalm 139:14).

Because of this, "almost Christian" doesn't cut it for the Lord. The Commandery is a well crafted counterfeit of Christianity, but that is all. The fact that it is a bright jewel of fake orthodoxy resting on a pillow of heresy is enough to make one doubt its authenticity. This little "Jesus ghetto" is nothing more than a token, a scrap from the vast pagan table of Freemasonry thrown to Christians.

A COUNTERFEIT CHRIST?

Pious appearances cannot be trusted. There are many cults which look Christian, and put on a show

71

of piety (i.e., the Jehovah's Witnesses). We need to see what kind of Jesus the Lodge professes and then compare that Jesus to the Biblical Messiah.

Jesus is mentioned in Masonic writings amid a crowd of other "great teachers" or "reformers."[13] He never stands out as anyone special, above others such as Buddha or Mohammed. That is not the Jesus of the Bible.

Even in the most authoritative ritual work, Freemasonry adamantly refuses to confess either the Lordship or the resurrection of Jesus Christ. For example, in the Maundy Thursday ritual of the 18° of the Scottish Rite, called the Rose Croix of Heredom, we find the following:

> We meet this day to commemorate the death of (Jesus), not as inspired or divine, for that is not for us to decide.[14]

Masonry does not regard Jesus as either divine or risen from the dead. By taking no stand, Masonry takes a stand. Trying to offend no one, it offends the One about whom it should be most concerned.

Albert Pike gave a teaching on the "fraternal supper" of the Scottish Rite, which is a kind of Lord's supper for the Prince of Mercy degree. It is a ghoulish and blasphemous mockery of Christ's institution, for he teaches:

> ...thus, in the bread we eat, and in the wine we drink tonight may enter into and form part of us the identical particles that once formed parts of the material bodies called Moses, Confucius, Plato, Socrates or Jesus of Nazareth. In the truest sense, we eat and drink the bodies of the dead...[15]

Pike is teaching that we are eating parts of the

dead body of Jesus! Sorry, but the Biblical Jesus is not dead! He is alive, and there are no dust motes of His physical body floating around the atmosphere for us to consume!

The resurrection of Christ is the central pillar of Christianity. Jesus arose from the dead as the Bible teaches, and Pike's writing is horrid and perverse heresy of the worst magnitude. He has denied the resurrection of Christ! He has called the Godhead of Christianity one of the Baalim, or false gods.[16]

Masonry collectively is adamant in its refusal to bow the knee and acknowledge Jesus as Almighty God. To do so would be to forsake their insipid and murky Great Architect for a vital relationship with Jesus in all His power and glory.

Like the New Age movement of which it is a precursor, Freemasonry denies the unique deity of Jesus so it can bestow deity upon us all. Eminent Masonic authors fall all over themselves to deny the deity of Christ. Clymer, a high level Mason, writes :

> ...in deifying Jesus, the whole humanity is bereft of Christos as an eternal potency within every human soul, a latent (embryonic) Christ in every man. In thus deifying one man (i.e., Jesus), they have orphaned the whole of humanity.[17]

Similarly, Masonic authority J. D. Buck claims:

> Theologians first made a fetish of the Impersonal, Omnipresent Divinity; and then tore the Christos from the hearts of all humanity in order to deify Jesus, that they might have a god-man peculiarly their own.[18]

Notice the meticulously chosen words, designed to convey condescension or even contempt for Christianity, words like "bereft" or "fetish." The latter term refers originally to a witch doctor's talisman, but has in recent years also come to mean an item found stimulating by sex perverts! It is evident that most of these Masonic authors care little for Jesus and even less for His disciples who insist on taking Him seriously.

This Masonic "party line" is very much in keeping with what witches like to teach about Jesus. When asked if Jesus was God, many of my witch friends would say, "Of course. Why should he be left out?" It is not surprising that there is a curious unanimity between Freemasons, witches and New Age devotees.

It is evident from all this that Masonry at best, neglects Jesus, and at worst, denies His deity and resurrection. That alone should be enough to determine the evil of this cult. But let us look at anther element by which we may judge the Lodge's Christian character. The Lord Jesus Himself warned:

> Not every one that saith unto me, Lord, Lord, shall enter into the Kingdom of heaven; but he that doeth the will of my Father which is in heaven.   Matthew 7:21

How well does the Lodge measure up to those standards?

7

Keeping God's Commandments

Freemasonry does a lot of promoting about how noble and godly it is, and many sincere people honestly believe it is a Christian organization.

In the Lodge of which I was a member, we did many nice things, including giving scholarships at high schools. Whenever we did, we always made certain the papers knew about it. It was good public relations.

"Masonry makes good men better," I used to tell my friends, and actually got a couple to join the Lodge. However, none of these noble deeds actually qualify the Lodge to be a Christian institution. Good deeds do not make an organization Christian, any more than they make an individual Christian (Ephesians 2:8-9).

LET'S ALL JOIN THE MAFIA!

A dear Christian lady brought her husband to our office to discuss the Lodge. He was a hard-bitten 32° Mason, but claimed to also be a Christian. I shared some things with him about the nature of Masonic belief in God and Jesus, but he was unconvinced.

He said he was a member of the Shrine, and had seen how many needy children were helped through the Shrine hospitals! He asked me how such wonderful things could be accomplished by an organization which was as pagan as I claimed.

I told him of large hospitals and churches in the city that had been built with money donated from Mafia leaders in the city who were Catholics. Much of this money was probably donated in a misguided effort to atone for their crimes.

I asked him if that made the Mafia a Christian organization. Should we all run over and sign up for the Mafia because of its good deeds? His ears turned red, and he said that no, we probably shouldn't.

I asked him if he realized that the Mafia, with its omerta (bloody secrecy oaths), its arcane symbols and its vendettas (blood feuds) was actually a creation of Freemasons. He didn't, of course, so I showed him evidence that Albert Pike's European colleague, Mazzini, was involved in the creation of the Mafia. The unfortunate fellow was finally willing to listen to reason and Biblical counsel.

A FLAWED DOCTRINE OF SALVATION

This shows the fallacy of Masonry pointing to its charitable foundations to prove its worthiness. This chapter briefly addresses whether the Lodge really is faithful to the commandments of God.

When Masons bring up their good works, they are following an anti-Biblical, religious view of how one pleases God. Sadly, it is an all too common view. They think God puts all your good deeds on one side of the scale, and all your sins on the other side. If you have more weight on the "good" side, you get to go to heaven (the Celestial Lodge). This doctrine is religion, but it is not Christianity!

The Bible teaches that sin is an offense against an infinite and holy God, and even one sin is enough to destroy your relationship with God. Even in human law, the quality of the person against whom one sins can make a difference in the punishment. A person who shoots a drug dealer who was shooting at him first will probably get a lighter sentence than a person who shoots a child on purpose.

The child is more innocent and harmless than an adult, fully-armed drug dealer. There is also something so viscerally heinous about killing a little child that it brings an almost universal outrage. God is infinitely more innocent than even the purest little child. Although human terms ultimately fail us when we seek to address God's qualities, He is such an unimaginably great and holy Being that a sin against Him bears a grievous weight far too great to calculate.

This means there is no such thing as a little sin. It also means that even if we only commit sin once in our lives (an impossible level of holiness), that one sin would weigh more than all the good deeds we could ever commit in our lives.

God's Law is perfect and integral (Psalm 19:7-8). Therefore, if we sin against one part of it, we sin against the entire law. If you shoot a person in the foot, you're charged with assault! Even though you shot only a relatively minor part of a human body, you are still charged with assault against the entire person. You are not given a lesser sentence for "toe assault." The Bible teaches that:

> ...whosoever shall keep the whole law, and
> yet offend in one point, he is guilty of all.
>
> James 2:10

This is why all the good deeds, all the hospitals or scholarships in the world cannot make up for one sin! Isaiah declares:

> But we are all as an unclean thing, and all
> our righteousnesses are as filthy rags; and we
> all do fade as a leaf; and our iniquities, like
> the wind, have taken us away.   Isaiah 64:6

Therefore, we cannot rely on a "good outweighs bad" approach. We must, in being true to the scriptures, conclude that "There is none righteous, no not one", (Romans 3:10) and that "All have sinned and come short of the glory of God" (Romans 3:23). Therefore, in looking at Masonry, we must carefully, but reluctantly, set aside all their charities. These are all "filthy rags" in God's sight.

BLOWING YOUR OWN HORN

At our ministry to Masons, we get a lot of mail from Masons who have been given our literature and are offended by it. One fellow wrote and enclosed a wad of news clippings from his local paper. Each contained some wonderful account of Masonic charity. There was about a dozen pictures of a Lodge

Master shaking hands with some person as he presented a check.

I wrote back to the fellow and asked if he realized how far his Lodge is from the commandments of Jesus Christ. He taught his disciples:

> Take heed that ye do not your alms before men, to be seen of them: otherwise ye have no reward of your Father which is in heaven. Therefore when thou doest thine alms, do not sound a trumpet before thee, as the hypocrites do in the synagogues and in the streets, that they may have glory of men. Verily I say unto you, They have their reward. But when thou doest alms, let not thy left hand know what thy right hand doeth: That thine alms may be in secret: and thy Father which seeth in secret himself shall reward thee openly.   Matthew 6:1-4

This is not a description of Masonic bodies. They make certain that everyone knows the great charitable deeds of the Lodge.

In His Sermon on the Mount, the Lord taught about doing good deeds:

> Let your light so shine before men, that they may see your good works, and glorify your Father which is in heaven.   Matthew 5:16

Freemasonry is definitely "blowing its own trumpet," contrary to these two teachings of Jesus! It doesn't glorify Jesus by what it does. It only glorifies itself. The Masons let their "light" shine only to attract members and defuse criticism.

No doubt, some Masons are giving their money for the best reasons. However, we must question their judgement. In giving money to an organization which refuses to glorify Jesus, they are pouring their resources down a rat hole! Their sacrifice is genuine, but they need to remember:

> Behold, to obey is better than sacrifice, and to hearken than the fat of rams. For rebellion is as the sin of witchcraft, and stubbornness is as iniquity and idolatry.   I Samuel 15:22-23

Remember these verses, for they will come back to haunt the Christian Mason who remains in his Lodge. God is neither pleased nor honored by sacrifice which is offered in a disobedient or rebellious fashion, no matter how great it might be. Masonry breaks the commandments of Jesus Christ in its attitude towards giving. And this is the least of its collective sins!

THE GREAT OMISSION

Jesus commanded His disciples:

> ...All power is given unto me in heaven and in earth. Go ye therefore, and teach all nations, baptizing them in the name of the Father, and of the Son, and of the Holy Ghost: Teaching them to observe all things whatsoever I have commanded you.
> Matthew 28:18-20

Christians are commanded to testify about Jesus to "all nations." This means winning souls to the Kingdom. Masonry prohibits this practice within the Lodge, thereby forbidding what Jesus commands!

Is this hard for the Christian Mason to believe? Let him try to witness Jesus to another Mason and see how far he gets before he is reprimanded for it. There may be Lodges where everyone is a Christian and this is tolerated, but they are not technically "regular and well-governed lodges." For example, the **Texas Ritual Monitor** (the ultimate source of Masonic dogma) forbids "all sectarian discussion within its lodge rooms" (p.89).

The Scottish Rite ritual monitor teaches:

> No man has the right to dictate to another in matters of belief or faith; no man can say that he has possession of truth as he has of chattel.[1]

Albert Pike taught that the Mason's creed holds that:

> No man has any right in any way to interfere with the religious belief of another.[2]

Let us look at what the "ancient" charges of Freemasonry, some of the most venerable and dogmatic teachings of the "Mother" Lodge in England have to say about this. These date back to 1723, and are part of the most authoritative beginnings of modern Freemasonry, which is felt to have begun in 1717:

> Though in ancient times, Masons were charged in every country to be of the religion of that country or nation, whatever it was, yet 'tis now thought more expedient only to oblige them to that religion in which all men agree, leaving their particular opinions to themselves.[3]

All these sources make it clear that Masonry has seen fit to overrule the commands of Christ, and to insist that those Masons who claim to be Christians disobey Jesus as a condition to participating in their rites. That is pretty bad, but things get even worse.

THOSE ABOMINABLE OATHS!

Perhaps no element in Blue Lodge Masonry is more problematic for a Christian than the issue of the oaths. Bible-believing Christians find two problems with these oaths. The first is that they exist. The second is that oaths involve the candidate submitting himself to horrible murder should he ever break those vows.

We must let the Bible interpret itself rather than allowing men to interpret it for us (II Peter 1:20). Although many Christians have weakened their stand on this point, Jesus clearly taught we are not to swear oaths of any sort. In fact, Jesus was so laboriously clear on this that He spends five verses on this commandment during the Sermon on the Mount:

> Again, ye have heard that it hath been said by them of old time, Thou shalt not forswear thyself, but shalt perform unto the Lord thine oaths: But I say unto you, Swear not at all; But let your communication be, Yea, yea; Nay, nay: for whatsoever is more than these cometh of evil. Matthew 5:33-37

This should cover the subject, but amazingly, countless Christian men swear oaths when they join the Lodge, totally oblivious. Lest they miss the commandment in Matthew, it is repeated:

> But above all things, my brethren, swear not, neither by heaven, neither by the earth, neither by any other oath: but let your yea be yea; and your nay, nay; lest ye fall into condemnation. James 5:12

Again, a pretty clear commandment against any kind of oath. James even prefaces it by saying, *"Above all things..."* That is pretty strong language.

82

The Masonic Order solemnly demands what Jesus forbids, just as it forbids what He commands. How can such an Order possibly be Christian, when it is continually at loggerheads with the Founder of the Christian faith?

A PIG IN THE POKE!

Suppose I ask you to buy a house from me, sight unseen. I assure you that it is a wonderful house, with thousands of square feet and all the trimmings. All I want for it is a million dollars cash. I also want you to give me the money up front without ever seeing the house, or even knowing that it exists, or that I own it in the first place.

Obviously, you would be foolish to take me up on such an offer and give me the money. No adult would buy a home without seeing it or determining that it was mine to sell. That would be buying a pig in a poke.

However, many men who might otherwise be astute businessmen, and would never buy a set of golf-clubs, much less a house, without thoroughly checking them out, are willing to gamble their eternal destiny on the spiritual equivalent of a "pig in the poke," the Lodge! You see, the Mason enters the Lodge supposedly in utter ignorance of what he is about to experience.

A CANDIDATE'S DILEMMA

The Christian candidate for Masonic initiation is asked to strip off all his clothes and any metal on him. He is blindfolded after being robed in a funny set of blue pajamas with a hole cut in the chest and the legs missing. A rope is placed around his neck

83

and he is led to the door of the Lodge. He knocks, and a challenge comes from within: "Who comes here?"

He is totally befuddled, so the fellow who is guiding him says for him:

> Mr.____, who has long been in darkness, and now seeks to be brought to light, and to receive a part in the rights and benefits of the worshipful Lodge, erected to God and dedicated to the holy Saints John, as all brothers and fellows have done before.[4]

Imagine a Christian, perhaps even a deacon or a pastor saying that! A Christian should have the "Light of the World," Jesus Christ, dwelling within his heart! (John 8:12) We are told in the first chapter of John's mighty gospel that in Jesus:

> ...was life; and the life was the light of men. And the light shineth in darkness; and the darkness comprehended it not. There was a man sent from God, whose name was John. The same came for a witness, to bear witness of the Light, that all men through him might believe. He was not that Light, but was sent to bear witness of that Light. That was the true Light, which lighteth every man that cometh into the world.   John 1:4-9

Note the irony, one might even say the nerve of Masonic ritual, to claim as its patrons John the Revelator and John the Baptist—one the author of these very words proclaiming Jesus the only "true light," and the other proclaimed by these words to be the witness of Jesus' light! Yet in this ritual, they force the Christian to say that he has "long been in darkness."

John says that Jesus' light shone into the darkness, and the darkness comprehended it not! For a Christian Mason to proclaim himself in darkness and beg "light" from Freemasonry is like someone who stands in brilliant noon day asking to borrow a flashlight with which to see! Such a man has implicitly, if not explicitly, denied the very Lord Who bought him (II Peter 2:1).

What "light" could Masonry possibly offer that Jesus has not already given to the Christian?

People at churches where I speak ask me how high a person has to go in Masonry before they realize it is not of God. I tell them it should be evident to anyone before they have even taken the very first oath in first degree!

After the candidate is admitted to the Lodge, still blindfolded and with a rope (cable-tow) around his neck, he is prayed for by the Master of the Lodge. He is led around the Lodge room and then caused to ritually approach the altar of Masonry. There he kneels, with one hand underneath the Holy Bible and a Square and Compass set (the chief Masonic icons), and the other hand resting on top of them.

The Master then leaves his throne in the East and approaches the still blindfolded candidate from the other side of the altar. He says:

> Mr._____, before you can be permitted to advance any farther in Masonry, it becomes my duty to inform you that you must take upon yourself a solemn oath or obligation, appertaining to this degree, which I, as Master of this Lodge, assure you will not materially interfere with the duty that you owe to your God, yourself, family, country or

neighbor. Are you willing to take such an oath?[5]

What would you say at this point? You are kneeling in an awkward position, half-naked, and blindfolded before an unknown number of total strangers. But you have just been offered the pig-in-the-poke. You have to take the Master's word that the oath you are about to take is as harmless as he claims. Because of the real intimidation involved in this situation, most men don't bother to critically reflect on their decision.

Some don't know what would happen if they refuse to take the oath. For all they know, they might be killed. After all, they have heard "those stories" about Masons.

RIDING THE GOAT?

Masons play on those anxieties, under the guise of horse play. I don't know if what I am about to describe is a calculated attempt on the part of Masons to defuse the criticism made upon the Lodge through ribald humor, or whether is it is simply "good-old-boy" fun. However, I had the "privilege" of sitting through initiations in several Lodges, and I observed them doing with every candidate what I am about to describe.

In my personal case, this odd sort of psychological warfare began when I was being interviewed as a candidate for the Lodge in my own kitchen. The two fellows who came asked my wife, Sharon, if it was alright if I joined the Lodge. She assented, and they assured her that they would take good care of me and see that nothing happened to me.

One of the men chuckled in a "hail-fellow-well-met" sort of way and clapped me on the back, saying, "We'll be sure to clean all the tar and feathers out of his hair before we send him home." Both of them laughed conspiratorially at great length, and Sharon looked at me as if to say, "Do you really want to join an organization with these creeps?"

This continued at the actual initiation, where I was kept waiting in an anteroom. The fellow who was in charge of keeping an eye on me said that I should not worry about riding the goat in the initiation—most guys did it and never fell off.

Another fellow came in and said that I shouldn't believe any of "those stories" about riding the goat. A third fellow winked at me and said they'd only lost a couple of candidates in the last year through death by violence, so I shouldn't worry.

It was all done in the manner of good-natured teasing. In the many initiations I observed or actually took part in, I saw a lot of variations of this kind of fraternity house humor. The only common thing in the many jokes and disturbing allusions was this business about riding the goat.

That is interesting when one recalls Albert Pike's teaching about the he-goat of the witches' sabbat, and the way witches in the Middle Ages demonstrated their allegiance to Satan. They had to consent to sexual intercourse with "the goat," (usually a high priest rigged up with a goat's head, but occasionally a real demonic form which looked goat-like). Or they had to perform the so-called osculum infamum (obscene kiss) which involved kissing the goat's backside to show their fealty to Satan.[6]

I did not have to ride a goat or get tarred and

feathered during my initiation, nor did anyone else. However, such jesting can unnerve a candidate to the point that he is highly reluctant to be uncooperative when confronted with the choice of refusing to take this mysterious oath.

Masonry is no laughing matter. But like the people who tell jokes about death and sex because both subjects frighten them, this jesting may be a form of dealing with the spiritual terrorism Masonry invokes. It may be a vestige, an echo of older days, before the masks of respectability had been donned, when Masons knew they did indeed have to submit to "the goat."

Under the circumstances in which he finds himself, the Masonic candidate has no viable option but to choose this pig-in-the-poke.

In the next chapter, we will examine those oaths, and see why they are so far from the bland benignity which the Master attempts to evoke in assuring the candidate.

8

Dangerous and Forbidden Oaths?

We left our intrepid candidate at the altar, trying to decide whether or not to take the oath. What the candidate does not know is that the Worshipful Master is deceiving him, by insuring the oath's harmlessness. He may not realize his deceit, but he is conveying a "truth" to the initiate which, as we shall see, is actually full of lies.

Most candidates are willing to proceed and take the oath. A detailed examination of the Lodge oaths alone are beyond the scope of this chapter, but here is the climax of the Entered Apprentice (EA) oath:

> All this I most solemnly and sincerely promise and swear, with a firm and steadfast resolution to perform the same...binding myself under no less penalty than that of having my throat cut across, my tongue torn out by its roots, and my body buried in the rough sands of the sea, at low-water mark,

where the tide ebbs and flows twice in twenty-four hours, should I ever knowingly violate this my Entered Apprentice obligation. So help me God and keep me steadfast in the due performance of the same.

(Worshipful Master:) In token of your sincerity, you will now detach your hands and kiss the book on which your hands rest, which is the Holy Bible.[1]

Here are the lies of the ritual! Obviously, having one's throat cut and tongue torn out would violate the candidate's duty to himself, to say nothing of his family, which loves and depends upon him.

More importantly, this oath violates the Christian's duty to God (as pointed out in the last chapter). These oaths are record breakers for violating the commandments of God!

If the candidate takes the oath seriously, he violates the sixth commandment (Exodus 20:13) "thou shalt not kill," by swearing to his own murder.

Some Masons say the oath is a sham and doesn't mean what it says. If so, that breaks the third commandment (Exodus 20:7), which forbids taking the name of the Lord in vain. He has invoked the name of God, and even kissed the Bible to insure his sincerity. If he didn't take the oath seriously, he took the name of the Lord in vain.

By swearing on the Square and Compasses, he has broken the second commandment against idolatry (Exodus 20:4) since those objects are given the position of graven images. They repose upon an altar of worship and are treated with extreme reverence by Masons. As we shall see from the mouths of Masonic authorities themselves, they are actually forms of the oldest idols in history.

As shown earlier, the god of Freemasonry is not the Biblical God, and by participating in the oath-swearing ceremony, the candidate has violated the first commandment (Exodus 20:2-3) to have no other gods besides the Lord God.

That is four of the Ten Commandments broken in one fell swoop!

YOUR DUTY TO COUNTRY

The Masonic oaths definitely interfere with the Mason's duty to his country! For example, in the third degree ritual, the candidate swears:

> I will keep a worthy brother Master Mason's secrets inviolable, when communicated to or received by me as such, murder and treason excepted.[2]

In the Royal Arch degree of the York Rite, even that small qualification is summarily removed. The candidate swears that:

> I will keep all the secrets of a Companion Royal Arch Mason (when communicated to me as such, or I knowing them to be such), without exceptions.[3]

At this degree, The candidate also swears that:

> I will not speak evil of a Companion Royal Arch Mason, behind his back nor before his face, but will appraise him of all approaching danger, if in my power.[4]

Finally, in the Royal Arch degree, the candidate promises to:

> ...employ a Companion Royal Arch Mason in preference to any other person of equal qualifications.[3]

These oaths can interfere with the Mason's duty to his nation. I heard several Masons boasting that during World War II, German Freemasons gave special treatment to U.S. POW's who were Masons. A few were even permitted to escape. That is nice for us Americans in retrospect, but those German Masons committed high treason.

Would we have wanted U.S. Masons doing the same thing to German Masons? I dare say not! Not only that, but the oaths mentioned above would also require some of the following very problematic scenarios:

1) An officer of the court who knew of an arrest warrant sworn out against a brother Mason would have to warn him immediately so he could flee the jurisdiction.

2) A Mason who was told of a brother Mason's crimes, even including rape, robbery, or child abuse, would have to keep his knowledge of those crimes a secret, even in a court of law!

3) A Royal Arch Mason who knew of a Companion Mason's being a murderer or a traitor would have to keep his knowledge a secret!

4) A Royal Arch Mason would be obligated to hire a Companion Mason, even for sensitive or skilled professions, even if he didn't have nearly the qualifications required.

Additionally, though not mentioned in the oaths, many times Masons get a "fairer" trial in courts where a Masonic judge presides. A sizeable majority of judges are Masons, and many attorneys are Masons as well. If a Mason appears in court against a non-Mason, all he has to do is give any number of obscure gestures or words to the judge, and the judge will be obliged to rule in his favor. No one in the court room will be the wiser (except another Mason, who would be forbidden from bringing the incident to light).

It is easy to see how these elements of the oaths could very definitely be detrimental to the welfare of our nation. Masons, it is said, "take care of their own," and they do, to an extent which is frightening.

It's obvious that our poor candidate has been sold a bill of goods! He has promised, in advance, to keep secrets he knows nothing about and to obey an oath which he has never heard. Can all this possibly be Godly?

A QUESTION OF TRUST

It is amazing that more Christian Masons don't see the light and get out early. Don't they see that they are the real temple, not the Masonic temple?

Their bodies have been paid for by the blood of Jesus and they have no right to surrender it to be hacked up by Masons. Their bodies are not their own (I Corinthians 6:20), and they should never allow Masons to touch what belongs to Jesus. The Christian's body is a temple of the Holy Ghost (I Corinthians 3:16) and it should be kept holy.

This is an essential spiritual truth. In Masonry's attempts to "build its temples in the hearts of men," it has forgotten that in the case of a Christian man, it is attempting to build where Jesus has already claimed the ground. In the ultimate sense, over and above all the commandment breaking and political double dealing, this is the most serious problem for a Christian Mason.

Most Masons who are Christians remain in the Lodge because of trust. Over and over again, Masons tell us, "My father was a Mason, my grandfather was a Mason! They were good men. If they were involved, it couldn't be wrong!" These Masons have chosen to trust men they respect rather than the Word of God.

The vast majority of Christian Freemasons join because someone they love or respect is a Mason. Often, it is a family member. They assume that this esteemed person has checked it out and found the Lodge to be perfectly alright, and that's good enough for them.

So when they see all the bizarre things like oaths and Christless religion, they tend to push it aside, saying to themselves, "If my dad (or whoever) is in this, it must be okay." They quench the Holy Spirit.

But they don't realize that their father joined for just the same reason, because someone he respected and trusted was a Mason. The vast majority of Masons have never bothered to examine what they are involved in. Thus, they have built their "temple of trust" upon a very flimsy foundation.

It is like these comic moments when the family goes on a picnic, and everyone assumes someone else brought the dessert. Everyone in the Lodge assumes someone else has checked Masonry out, and it is alright. Unfortunately, 99% of Masons are ignorant of their own Craft. They are too busy, or too lazy to investigate it for themselves.

None of these men can bring their fathers to stand with them before the judgement bar of Christ and say, "I joined the Lodge because of him, it's his fault." Jesus expects us to take responsibility for our actions. He expects us to trust in Him and His Word, even when it is hard:

> My people are destroyed for lack of knowledge: because thou hast rejected knowledge, I will also reject thee, that thou shalt be no priest to me: seeing thou hast forgotten the law of thy God, I will also forget thy children.   Hosea 4:6

Each of these men are a link in a chain of trust. Unfortunately, that chain is ultimately forged to a ball of iron which is about to be thrown into the lake of fire!

It is tragic that more of them did not have the discernment or courage they needed. They didn't understand the cost of the "devil's bargain" they made at the Lodge altar. Paul wrote:

> Know ye not that ye are the temple of God, and that the Spirit of God dwelleth in you? If any man defile the temple of God, him shall God destroy; for the temple of God is holy, which temple ye are. Let no man deceive himself. If any man among you seemeth to be wise in this world, let him become a fool, that he may be wise. For the wisdom of this world is foolishness with God. For it is written, He taketh the wise in their own craftiness. And again, The Lord knoweth the thoughts of the wise, that they are vain.
> I Corinthians 3:16-20

The Masons boast of their "Craft," and indeed, worldly wisdom is imparted through it. However, is not swearing an oath to permit the destruction of a Christian's body (the temple of the Holy Spirit) one of the ultimate forms of defilement?

The Lord Jesus continually brings down the wisdom of the wise and confounds it as utter foolishness before His mighty Word and power! Many Masons are doing what I did, renouncing the Lodge. We have become fools, so that we may be truly wise!

He who dwells in the body of any Christian is far, far greater than all the Masonic temples in the world. Who, except Satan, would dare to say they could improve what Jesus has already cleansed? The Lord warned Peter, "What God has cleansed, that call not thou common" (Acts 10:15).

9

The "Eastern Star"

Next to Freemasonry itself, our ministry is most questioned about the Order of the Eastern Star (OES), so we will cover some background dealing specifically with the "Star" (as it is fondly called).

The Order of the Eastern Star was founded in 1868 as a women's auxiliary for the Lodge. It is open to all female relatives of Masons, and functions under the authority of the Lodge. A Master Mason, called the Worthy Patron MUST be present at all Star meetings. There is also another Mason, an Associate Patron, usually present. Otherwise, all the offices are held by women.

To ask what the Star does, is a bit like asking what a Lodge does. The easy answer is: Not much. However, the Star does what little it does with a tremendous amount of pomp and ceremony.

Star meetings, like Lodge meetings, consist of opening the meeting, with much waving of rods and banners and the singing of hymns. The officers ritually declare their stations and functions and Grand Chapter officers are introduced and honored with tedious predictability. This can easily kill a half hour to forty-five minutes.

Then, minutes are read, sick members are mentioned and any items of mundane business transacted, much like any other club. If members are to be initiated, that is done; and that can take at least an hour.

Then the chapter is solemnly closed, with a lot more ceremonial ado, followed by a social hour. ULTIMATELY, the chief function of a Star chapter, like a Lodge, is to make more members! Everything else is secondary to that.

What makes the Star unique is its feminine character and its use of a large five-pointed star and five women Bible characters as key ritual features while the initiate is led through its "Labryinth". These five "heroines" of the Bible are Jephthah's Daughter, Ruth, Esther, Martha and Electa. As the candidate is led through the Labryinth, each "Point" officer teaches them a pious lesson based on the life of the woman they represent.

In light of the huge amount of oath-swearing in Masonry, it is instructive to note that of the many Biblical heroines, the rather odd character of Jephthah's daughter is the first introduced. Someone who put this thing together had a grisly sense of humor, since she is featured in the Bible as an innocent daughter who was literally sacrificed because of her father's fidelity to a rash vow (Judges 12:29-40).

That is a perfect metaphor of what happens

spiritually to women in Masonic families!

The Star is regarded by Masonic women as a fine Christian institution within Masonry. I was an Associate Patron in a chapter and can see how this might be assumed.

Classics like *How Great Thou Art* are sung. The motto of the chapter is right out of the Bible: "We have seen his star in the east, and are come to worship him" (Matthew 2:2). Two of the "points" of the "Star" are from the New Testament.

For all this piety, the Star is actually one of the most blatant examples of satanic gall in all Masonry! The symbol of the Star is an inverted, five pointed star, known in witchcraft as a pentagram.[1]

The pentagram is regarded as a very powerful magical device, and is perhaps the best known symbol of witchcraft! Its association with witchcraft is undeniable and is the most common symbol of Satanism in the world today.

The inverted pentagram is the official symbol of the two largest satanic churches, the Church of Satan and the Temple of Set. This inverted star, with the goat's head within it (called "Baphomet") is on the cover of The Satanic Bible![2] You can see it in satanic graffiti and ritual sites! It is also found on the albums of satanic rock groups like Venom's *Black Metal* [3]or Slayer's album, *Hell Awaits*. [4]

The association of the pentagram (especially inverted) with witches, magic and evil, is much more ancient than the Eastern Star.[5] The pentagram has appeared in magic texts for centuries, and may predate the time of Christ.

Why does a supposedly Christian organization have the ultimate symbol of devil worship as its logo?

PLAYING WITH WORDS

To answer that question, we need to understand what the "Eastern Star" actually means. Sadly, one of the key strategies a cult uses is to play upon the common interpretation of words. Dealing with any cult involves getting through a jungle of definitions.

The cult will use words which have common meanings, like "Jesus" or "saved," or well known Bible verses, but will carefully not explain to potential members that they have applied a subtext to these terms, a second layer of meaning!

This is the problem with the OES. The people who put the order together back in the 19th century relied upon the familiarity with Matthew 2:2 as a common "Christmas verse." However, in light of the satanic symbolism involved in the OES, we need to look for another layer of meaning applied to that verse of scripture.

A STAR IN THE EAST?

The A.V. 1611 text does not say "eastern star." It says "star in the east." An examination of the text reveals that we are being exposed to a verbal scam. Since the wise men were from the Orient (i.e., Persia), the star which they saw over Bethlehem could not have appeared eastern to them, but western.[5]

This text means the wise men were in the east when they saw the star. Then they headed west to Bethlehem! You see, there is an important reason why the OES is called Order of the Eastern Star, and sadly, it has nothing to do with Matthew 2:2!

The phrase "Eastern Star" has a specialized meaning in occultism. It refers to the star, Sirius,[6] which is the most significant star in Satanism! It is sacred to the god, Set.[7] Remember Set as the evil Egyptian god who killed Osiris? Set is probably the oldest form of Satan! The Eastern Star is the star of Set.

Thus, when the OES uses Matthew 2:2 "We have seen his star in the east, and are come to worship him", they rely upon all these nice ladies assuming that "his" refers to Jesus.

However, deadly word games are being played here. Though most women involved in the OES doubtlessly assume they are worshiping Jesus as they kneel around a huge satanic pentagram, it is obvious that the "his" actually refers to Set's star, not Jesus' star.

The heritage of Sirius is so central to Satanism that we need to examine its place in Masonry. This subject alone should be enough to expose the satanic roots of the Lodge.

THE BLAZING STAR"

Is it a coincidence that the inverted pentagram is used as the signet of the OES? From its monitors and authorities we learn about the profound reverence due the pentagram in Masonry. For example, in the first degree, we learn about Masonic "Ornaments." The new Mason is taught that:

> The ornaments of a Lodge are the Mosaic pavement, the indented tessel, and the blazing star...Divine Providence...is hiero-

glyphically represented by the blazing star in the center.[8]

The center of the Lodge is a "blazing star" which supposedly symbolizes Divine Providence. The illustration on the next page shows the prominent place given the pentagram in the Lodge room.

However, we can dig yet deeper into the meaning of this star. In Albert Pike's commentary on this degree, we find the usual duplicity found elsewhere in the Lodge. He explains:

> To find in the BLAZING STAR of five points an allusion to Divine Providence is fanciful; and to make it commemorative of the Star that is said to have guided the Magi, is to give it a meaning comparatively modern. Originally, it represented Sirius, or the Dog-star, the forerunner of the inundation of the Nile...Then it became the image of Horus, the son of Osiris, himself symbolized also by the Sun, the author of the Seasons and the God of Time...It became the sacred and potent sign or character of the Magi, the PENTALPHA...[9]

Pike readily casts aside the bland lie of the degree and confirms that the blazing star is neither Divine Providence, nor is it Jesus' "star in the east." It is an Egyptian idol, the symbol of Sirius!

Sirius is magically regarded as the most dangerous star in the sky. The Egyptian people suffered the most during its time of ascendancy. It reached its apogee in the Egyptian sky on July 23. This was the hottest, driest time of year for the civilization, when the Nile was at its lowest ebb—the Nile, upon which Egypt depended for irrigation.

EAST – Worshipful Master

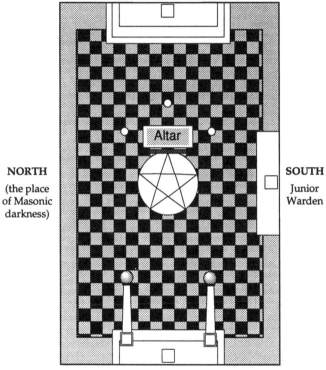

NORTH

(the place
of Masonic
darkness)

Altar

SOUTH

Junior
Warden

WEST
Senior Warden

The Temple of Blue Lodge Masonry

Thus, Sirius was a star of scorching, blasting evil. It was the most dreaded omen in the heavens.

Its association with the dog or hyena, is ancient. Oddly enough, we carry in our modern language a reference to this scorched time of year. The time of great heat and humidity from mid-July to mid-August is often called the "Dog-days." The reference is to the Dog-star, Sirius.

In identifying Sirius, we have come very close to identifying the true deity of Masonry by yet another of his many masks.

SPIRITUAL ADOPTION

There is another critical problem in the Star—a reason why no Christian woman would wish to be a part of this organization. The Star is known as "Adoptive" Masonry. This means that mothers, daughters, wives and sisters of Freemasons are spiritually "adopted" into the Masonic order, even though they aren't actually Masons. Hence, they submit themselves to Masonry's spiritual authority.

Set (Lucifer) is the acknowledged god of Masonry. Thus, what does it mean for a Christian woman to be adopted into the Star? She is submitting herself to the spiritual authority of Lucifer.

She may not know it, but in bowing before the altar of the Star, the inverted pentagram of Baphomet, she has surrendered herself (however innocently) to the gods of Masonry. That WILL give Satan an entry point into her life, no matter how devout a Christian she may be.

A Christian woman is already adopted into

God's family (Romans 8:15-17), but what happens to her when she has unwittingly entered the temples of Lucifer and surrendered herself to his power? At the very least, it creates incredible inner turmoil. She has become a member of two families—families which are eternally at war with each other.

If she is a very strong Christian, she may be able to weather such an internal civil war, but she must understand that this conflict will be passed on down to her much more vulnerable children! (Exodus 20:5)

Think about it, ladies. Look at the spiritual and moral consequences of the involvement of thousands of good, Protestant women in the Star between the end of World War II and today. The children raised by those women, those born into the "Baby Boom" generation, have given themselves over to occultism, immorality and evil on a scale unprecedented in this nation's history! Why?

Why have men and women born in "good, Christian homes" between 1945 and the present been given so many evils? Abortion runs rampant, the "hippie" drug culture almost wasted an entire generation, occultism and witchcraft are now in every bookstore.

All these horrible things can be traced, I believe at least in part, to the fact that the women who raised this generation were the first women to give themselves with any degree of enthusiasm to the Star.

With the Star, Lucifer has driven a dagger right into the heart of American motherhood! He has taken women who would never be caught dead doing anything witchy or immoral, and has beguiled them with a pious sounding cult into worshiping the gods of child sacrifice and sexual promiscuity. Their children are reaping the grim harvest of Lucifer, or Set, on those very sinful practices!

AN "IPSISSIMUS" SPEAKS

What does the "Eastern Star's" god actually represent? Dr. Michael Aquino is the leader of the Temple of Set, an elite satanic church. Aquino calls himself an "Ipsissimus," a term taken from the Golden Dawn, an influential Masonic/Rosicrucian society which served as the incubator for some of the premier magicians of this century, including S. L. Mathers, Dr. Wynn Westcott, Aleister Crowley and Arthur E. Waite. All these men were also high level Freemasons!

The title, Ipsissimus, is the highest to which one may attain, and it implies that you are a god on earth. Few people take the title, even within the ego-maniacal world of Satanism. Aquino did, and has emerged as the most articulate spokesman for the satanic movement today.

Aquino claims to have had visions of Set, and to have been anointed as a kind of "prophet" to continue the path begun by Aleister Crowley prior to the latter's death in 1947.[10] Giving this man's credentials was necessary to show his authority to speak on the subject of Set.

Aquino has written that Set is the oldest formulation of the being now called Satan, the embodiment of the sense of alienation and loneliness which man feels from the universe.[11] Man, Aquino claims, is in a unique state of separateness from the rest of creation and feels terribly alone. Set represents that reality of

106

man's fractured relationship with his universe. That is a limited, but workable secular definition of "sin."

This is what the "blazing star" at the heart of every Masonic lodge represents: the Dog-star, Sirius—the symbol of Set! It represents the profound breach between man and God.

In other words, it represents sin. Though Aquino may not believe in sin, he has put a name to sin's pain, and that name is Set! This is why the pentagram, the symbol of Witchcraft, is also a prominent part of Masonry and the Order of the Eastern Star. They have the same god—Satan, or Set!

10

Kindergartens for Satanism?

Having seen the peril of Freemasonry, the question immediately arises: "What about the youth orders of Masonry? Are they THAT dangerous?"

The main ones are the Order of DeMolay for boys and the orders of the Rainbow Girls and Job's Daughters for young women.

In some respects, these organizations are like the Boy or Girl Scouts. And they provide allegedly innocent fun and fellowship. However, parents who send their son or daughter to one of these Masonic youth groups assuming they will be exposed to only wholesome influences are sadly mistaken.

The DeMolays are an order in which the central ritual drama revolves around the trial of Jacques DeMolay, the last Grand Master of the Knights Templar. The ritual is intended to teach loyalty, honor and fidelity to promises or obligations.

The Job's Daughters use the Biblical characters of

Job and his daughters to teach similar pious virtues, but of a more maidenly character. Rainbow is even more bland than the Job's Daughters, and little seems to be out of place in them.

However, their very innocuous qualities should send up a warning flag, as they are so closely associated to the adult Masonic orders.

SPIRITUAL AUTHORITY

That very close tie is the most critical concern because it boils down to a question of spiritual authority. Hopefully no Christian parent needs to be reminded of the responsibility they have for the precious young souls that God has entrusted to them (Deuteronomy 6:4-7, Ephesians 6:4).

Additionally, Jesus Himself warned about those who lead their children into error, that:

> It were better for him that a millstone were hanged about his neck, and that he were drowned in the depth of the sea.
>
> Matthew 18:6

Parents have spiritual authority over their children; and if they are Christians, the Lord has placed a special hedge of protection about the family (Job 1:10). Only a few things can break that barrier down and allow Satan access. One of them is the parents' involvement in idolatry, permitting their children to worship at strange altars!

This is a spiritual principle which has little to do with the appearances of these youth groups. Most of them, except the DeMolays, are quite harmless aside from teaching a kind of salvation through good deeds (which runs through all of Masonry).

109

However, all these groups are chartered under Masonic authority. NONE is allowed to meet or function without the presence of a Master Mason. No one can join unless related to a Master Mason.

These groups are "feeder programs," designed to draw young people into the adult Masonic orders. They are technically called "adoptive" Masonry, and the youth who joins is spiritually adopted into the Masonic family, even though not actually a Mason.

The SPIRIT of Freemasonry looms over every meeting of DeMolays or Rainbows! Whether he knows it or not, the Mason who is present brings with him a luciferian priesthood, which envelopes all who are involved in the meeting.

Even a Christian youth who has been adopted into the family of God (Romans 8:15-17) comes under this spiritual authority if he or she joins one of these orders. This cannot help but make them "double-minded" (James 1:5-8) at an already difficult phase of their lives.

No matter how innocent these ceremonies appear, they function under the shadow of an anti-Christ spiritual power which the Master Mason brings with him.

WOULD YOU LET YOUR CHILD ATTEND A WITCH COVEN?

The very question is ludicrous! Yet, a witch coven meeting might be less dangerous than a Rainbow meeting! At least the Christian girl would be on her guard if she knew she was at a witches' sabbat! She would be praying her little heart out under her breath and would be wary of everything that was said or done.

110

The same spiritual principality which presides over witchcraft also reigns over all Masonic youth orders! With Masonry being based on sexual fertility cults, the young person walks innocently into a spiritual minefield without warning or advance knowledge when joining a Masonic youth order. They stand in Satan's temple, surrounded by insipid ceremonials which give no warning about the danger lurking beneath them.

Unlike the child at a witches' gathering, they have no reason to be on guard. Their parents sent them there! Their friends are all around them. Grown-ups are present! Everything SHOULD be fine, but they have opened a doorway for Satan to begin corroding their very souls! They have broken down the hedge which God has around them, often with the help of their parents! (Ecclesiastes 10:8).

At the age when powerful hormones are running at full capacity, these young people are surrendering themselves unknowingly to a spirituality which is engineered to provoke lust!

The rites and symbols of Masonry are sexual in spirit. A young person who is receiving moral doctrine at home is having that teaching subtly undermined by submitting their spiritual priestcraft to the Master Mason who's in charge!

Masonry exalts sexuality to the level of deity, but in a disguised, allegorical fashion. As mentioned earlier, the square and compasses and other Masonic symbols are veiled references to the human reproductive organs—talismans designed to increase sexual desires.

While these orders teach pious principles on the

surface, they are pouring fuel on the smoldering fires of teen emotions. The God given rights of the parent are diluted by the very real authority of a strange god who appeals to all that is carnal and fleshly in the young person.

POMP AND CIRCUMSTANCE

While we are on the subject of carnality, let us look at a less dangerous, but still troublesome quality of these youth groups. They tend, by their very nature, to make the child prideful. The desire to be something "special," and to lord it over their peers is a common adolescent itch. It is not something Christian parents should encourage.

The youth orders do much to "scratch" this carnal desire to be "some great one" (Acts 8:9). The young people get to feel special because they belong to a secret club. Those who become officers are given typically impressive sounding titles: Honored Queen, Master Counselor.

They are dressed in all the pomp of royalty—crowns, formal gowns, satin cloaks, and chains of office. All these serve to glorify their egos instead of conforming them to the humble image of Jesus (Philippians 2:5-9, Romans 12:1-2).

These orders make a great show of teaching religion, but that is NOT what our young people need! They need Jesus, and His gospel of grace! Subtly, but completely, a doctrine of salvation by works is taught in each of these orders.

The secrecy and elitism is also contrary to true Christianity, which does not exclude, but rather embraces whosoever will come.

THE "JACK THE RIPPER" MEMORIAL HOME
FOR BATTERED WOMEN?

There is one youth order which stands out above the others in terms of its audacious embrace of evil. That is the DeMolay order for boys. Though all the youth orders share the dangers mentioned above, DeMolays are especially dangerous because they serve as the incubator for future Masons.

Although I never was in the DeMolays, I have a dear friend and colleague in the ministry who was, but is now saved by Jesus. From his personal experience as a member, he has referred to the DeMolays as a "kindergarten for Satanism." He believes it was a major stepping stone for him into occultism and witchcraft.

It is an especially grisly jest to name the Masonic order for young men after Jacques DeMolay, the last Grand Master of the Templars.

The DeMolay ritual makes a great hero of its namesake. He is held up as a paragon of manly loyalty and virtue. What the order's ritual does not tell its young charges is that DeMolay was burned at the stake for being a homosexual, a pedophile (lover of young boys), and for practicing witchcraft and worshiping a false god named Baphomet! (See chapter 15 for more on the Templars' history.)

Naming a boys' order after DeMolay is like naming a shelter for battered woman after Jack the Ripper, or a home for unwed mothers after serial killer Ted Bundy! Though the Templars' history is controversial, DeMolay died cursing those who put him to death—hardly a model of young Christian manhood!

Why, with all the great men in western history (including Jesus!), would the Masons pick such a corrupt, controversial and obscene man to be a role model for their young men?

Why not Stephen, the first martyr? Why not Joshua? The answer is because DeMolay is one of the central "idols" of the Freemasonic pantheon, probably second in stature only to Hiram Abiff. The DeMolay ritual prepares the young Mason-to-be for a life of involvement in societies dedicated to the worship of Baphomet!

PROTECT YOUR KIDS!

Satan knows that many Christian parents are desperately seeking good places to send their children for fellowship. He has flanked them on every side with filthy movies, heavy metal music, pornography, drugs and gangs. Parents need to be especially aware of the more subtle attacks on their young people.

Masonic youth orders are poison in unlabeled bottles. They are even more dangerous because they are often endorsed by uninformed pastors.

Many heartsick parents have written our ministry saying they allowed their child to join DeMolay or Job's Daughters because their pastor devoted a special Sunday service to them and allowed them to show up in all their glorious regalia in church for the entire congregation to admire. What parent wouldn't trust their child to something endorsed by their own pastor?

Masonic orders are dropping in membership, and they are desperately seeking new young people. The PRIME reason these orders exist is to pump young adults into dying Masonic organizations—to acclimate them to exotic and pretentious rituals and secret oaths, both emotionally and spiritually! Although most Masons don't realize it, they also exist to ensnare the young men and women in the delicate web of demonic oppression.

Not all kids who join these youth groups become drooling demoniacs, anymore than all children who smoke a joint end up crack addicts! However, not an insignificant number of them do go on to dabble in occultism or witchcraft! They are being exposed unnecessarily to serious spiritual danger in joining such groups!

If your child has friends whose folks are Masons, be on the lookout for attempts to draw them into these orders. Their social activities seem harmless, but are used to draw kids into their web.

Watch for your own peripheral relatives (cousins, etc.) who are Masons and may be working on your children. In communities (or churches) where the Lodge is very strong, peer pressure is a real factor also. Everyone who is "anyone" in high school will be in DeMolay or Rainbow. Kids find it extremely difficult to resist that kind of pressure.

A child cannot join these groups without the parents' express permission (in most states), so you must take a firm stand. If your child asks about joining, take time to discuss it openly. Treat it as if that child were thinking of joining the Mormons or the Moonies!

COMMUNICATE WITH THEM!

Be certain your lines of communication are open at all times. Make certain your children know they can come to you at any time, with any problem, without being condemned! That is so very important!

Tell your child, in simple terms, what is in this chapter. Make it clear that people who are in these organizations are not evil, but deceived victims. Help them understand the danger is no less real, just because it is invisible. Help them see that Masonry itself is a huge and dangerous cult, but that its members are victims. In most cases, forewarned is forearmed!

The best, most important thing you can do for your children is intercede for them daily! Plead the blood of Jesus over them every time you think of it throughout the day. Teach them to be prayer warriors themselves! The best defense is a good offense!

PASSING YOUR SEED THROUGH THE FIRES!

Failing to protect your child from these groups is, in a spiritual sense, what the people of Israel were condemned for doing in Jeremiah 32:34-35. You would be passing your seed (children) through the fires of Molech, a pagan fertility idol.

In the days of Israel, this amounted to literal human sacrifice. Today, the tender young souls of our teens are being passed through the fires of occultism and Lodgery!

In our society, there are enough "fires" to walk through: drugs, sex, horror videos and rock music. But in the Masonic youth groups, the young person is often urged to join them with the "blessing" of parent or even pastor. Solomon warned that "He that troubleth his own house shall inherit the wind"

116

(Proverbs 11:29). His own life was tragic proof of that! His idolatry fractured the kingdom of Israel, and his children did indeed inherit the wind.

We trouble our own houses when we permit our children to join these pagan societies. Subtle spiritual discord is created in their lives when they sit, week after week, under the dark penumbra of Freemasonry, and yet we expect them to be good and upright adolescents.

Let us remember Jesus' "curse of the millstone" (Matthew 18:6) and protect our young people from passing through any more fires than our culture already exposes them to!

11

Masonic Trinkets, Tokens and Trouble!

Part of the fascination and the danger of Masonry is in its use of symbols and tokens (ritual gestures).

The danger comes first from the fact that many of these things are occult; and secondly, they are used to perpetuate a "good-old-boy" system of favoritism in business and government. This is how Masons recognize one another without a word being spoken.

First, there are the famous rings, tie tacks and lapel pins. Some of these are obvious, like the Square and Compass. They symbolize the human reproductive organs, locked in coitus (when displayed together). Others are a bit more subtle, and some are quite fiendish.

In the York Rite, there are lesser known symbols in the jewelry. One is the keystone, which is from the three lower degrees. It is supposedly "the stone which the builders rejected," which became "the headstone of the corner" (Psalm 118:22). It is from the Mark Master (fourth degree) and represents Hiram Abif, the Masonic "Christ figure," who supposedly carved the keystone.[1]

It is a trapezoidal stone with a double circle engraved within it. Inside the circle are the letters: H.T.W.S.S.T.K.S. This stands for **Hiram, the widow's son, sent to King Solomon.**[2] (See Illustration 1.) This entire concept denigrates Jesus in favor of the Masonic hero, Hiram. The Bible makes it clear that Jesus is the stone referred to in Psalm 118... (See Matthew 21:42, Mark 12:10.)

1. York Rite Keystone (Mark Master)

2. Royal Arch Tau Cross

Another common York Rite symbol is the Tau cross. (See Illustration 2.) This distinctive cross, which often looks like the letter "T," is actually the symbol of the pagan slain and risen god, Tammuz (Ezekiel 8:13-14).[3] It is a symbol for just another counterfeit Masonic "Christ."

The other York Rite jewelry you may see is the Templar symbol. (See Illustration 3.) It is a large Maltese cross with a circle in the center. Inside the circle is a red Latin cross within a crown. Around the arms of the cross is the Commandery motto, "In

119

Hoc Signo Vinces." (In this sign, conquer!)[4]

Although this may seem harmless enough, the motto is originally attributed to the emperor Constantine, who used it in conjunction with a supposedly heavenly vision to begin the subversion and politicization of Biblical Christianity into the false, apostate Alexandrian cult.[5]

3. Templar Symbol

A similar shell game is played with the word, "sign," as in the Eastern Star with the phrase, "his star in the east." The sign Constantine referred to was NOT a Christian cross, but a kind of "X" which had both Christian and pagan associations.[6] In modern magic, it is the sign of the slain and risen Egyptian god, Osiris[7] (another version of the "slain and risen" Hiram Abif).

Again Masonry has downgraded Jesus and replaced Him with its own "christ."

THE "BAPHOMET CROSS"

4. Scottish Rite, or "Baphomet" Cross

A Masonic symbol seen less frequently is the 33°
cross because it appertains only to the highest
degrees. It is more commonly called the Crusader's
Cross or the Jerusalem Cross. (See Illustration 4.) It
was supposedly worn by the first Grand Master of
the Knights Templar, Godfrey de Bouillon, after he
liberated Jerusalem from the Muslims.[8]

This symbol is on the hat of the Sovereign Grand
Commander of all 33° Masons in a very slightly
modified form. It is part of the magical signature of
Aleister Crowley, the supreme satanist of this
century![9] It is also found as the logo of the new
Catholic Bible, the Jerusalem Bible![10]

THE SIGN OF THE CRESCENT

Another more common, yet sinister jewel
Masons wear is the Shriner pin. (See Illustration 5.)
Although this comes in several styles, it usually is
some form of a crescent combined with a scimitar.
This identifies the wearer as a Shriner, either a 32°
Mason or a Knight Templar.

Aside from its witchcraft associations, the

crescent is the sacred symbol of Islam,[11] a false religion and one of the most serious rivals to Christianity! The scimitar is a sword developed by the Islamic armies. It spreads the "gospel" of Islam, that "There is no god but Allah, and Muhammed is his prophet."

5. Shriner Pin

Instead of the usual preaching techniques, the Muslims had a unique approach. It was called jihad, or "Holy War." The scimitar is the ultimate symbol of the jihad. This sword was used to cut off the heads of those who refused to bow down to Allah, including millions of Christians over the centuries! Would you want to wear on your lapel a symbol of the slaughter of countless Christians?

PHOENIX RISING

In the Scottish Rite, the premier symbol is the double-headed eagle. This is the most common icon of 32°. Although it has associations with 18th century Prussian royalty, most Mason authorities agree it is ultimately a symbol of a mythic bird known as the Phoenix. The eagle is also associated by Albert Pike with the Egyptian god Mendes.[12] (Remember the satanic "Goat of Mendes"?)

This bird (called the Bennu by ancient Egyptians) is said to live 500 years, burn itself to ashes on a funeral pyre, and then rise from the ashes in new youth to live another cycle of 500 years. It is a common symbol of reincarnation (an occult and witchcraft doctrine) and immortality.[13] This is a reflection of the Mason's belief in immortality (but AGAIN, without Jesus!). The Phoenix is a false symbol of a false and Christless "resurrection," the resurrection of the damned (John 5:29)!

The two heads on the bird, pointing in opposite directions, represent the Masonic (Gnostic) doctrine of the necessity of both good and evil, light and darkness,[14] which is dualism—a dogma contrary to the Bible.[15]

Manly P. Hall writes that this eagle represents the "Supreme Hierophant (High Priest)," a perfect being "in which all opposites are reconciled."[16]

He states that the only man who may wear this symbol is "one who is born again and has approached the throne of divinity. He is more than a man yet less than a god; therefore he is a god-man."[17] This obviously is NOT referring to the Christian experience. It is Masonry's pagan substitute—becoming a "god-man."

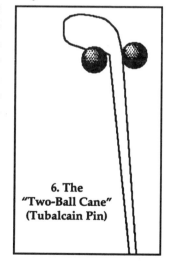

6. The "Two-Ball Cane" (Tubalcain Pin)

A VERY DIRTY JOKE!

For Masons who wish to conceal their member- ship from non-Masons, but still advertise it to their Lodge brothers, there is a special pin (or tie tack) they can wear. It looks like an upside down golf club with two balls near the top. (See Illustration 6.) Many people assume the person is a golfing enthusiast, but it is actually a visual Masonic pun.

This is called the "Two Ball Cane," and is a pun on the secret password of a Master Mason, "Tubalcain." It indicates that the man wearing the pin is a Master Mason, but it would not be recognized by a non-Mason. It is also an all-too-obvious pun on the "god" of Masonry, the male reproductive organ! Nice, eh?...especially when many men wear these wretched things to church on Sunday!

THAT APRON!

The apron is one of the best known identifications of a Mason. It is usually not worn in public, except for Masonic events like open installations of officers (a major recruiting tool), cornerstone layings, or funerals. There are actually many kinds of aprons, but we will only mention a few of them.

Masons receive a special plain white "lambskin" apron when they attain their Master Mason degree. Nowadays, they are usually lined with plastic to keep them clean, and are seldom worn until the Mason's death and burial.

At meetings, most Masons put on a white, cotton apron from a box at the door of the Lodge room. It is a generic, white, all-purpose apron worn by the vast majority of Masons unless they go up the ladder of degrees or become Lodge officers.

Officer aprons are fancy, and are often made of leather and trimmed in blue velvet with the "tool" (square, plumb, level, etc.) pertaining to the office embroidered upon them. Grand Lodge officers' aprons are even fancier, with silver tassels and balls.

From state to state, the officers' aprons are different, but they all contain occult symbols, like all-seeing eyes, the twin pillars, and the square and compasses. A Past Master of a Lodge also gets a special apron which he can keep. His apron has a picture of the sun-god, Ba'al, surround by a compass.

The older officer's aprons (from 1800-1900) were incredibly ornate, and quite beautiful, but crawling with magical symbolism. Even the simplest Masonic apron is designed to be an occult tool.[18]

Although the Mason is never told this, it goes back to the fig leaf apron which Adam and Eve made in the garden of Eden to cover their nakedness.

God cast aside those aprons (Genesis 3:21) because they were symbols of man's attempt to make restitution for his own sins. Instead, God killed an animal, thus shedding blood, and made them coats of skin. He was teaching Adam and Eve that without the shedding of blood there is no covering for sin. The Masonic apron, just like its fig leaf ancestor, is an attempt to walk around that important spiritual reality.

Actually, in temple Mormonism, Satanism and ceremonial magick, aprons are the symbols of the priesthood of Lucifer.[19] Satan has Masons wearing the skin of the lamb and trusting in it, instead of trusting in the blood of THE LAMB! What mockery!

SIGNS AND TOKENS

Masons may identify themselves through cryptic words or gestures that are too numerous to catalog completely. However, it may be helpful to know some basic ones. First, Masons usually identify themselves by shaking hands. A pressure with the thumb on the space between the second and third knuckle of the other person's hand usually is sufficient to identify oneself as a Master Mason.[20]

When a handshake is not possible (as in a courtroom), a Mason might approach the bench by the:

> "three upright regular steps...(stepping) off with your left foot one full step, and bring the heel of your right to the hollow of your left foot; now step off with your right foot, and bring the heel of the left in the hollow of your right foot; then step off with your left foot, and bring both heels together."[21]

It sounds complicated, but any Masonic officer has done it a thousand times, and can make it look as natural as can be.

Another way is through phrases, either simple or complex, depending on the circumstances. For example, a Masonic defendant in court might say, "I was hoping to get a SQUARE deal, your honor," with just a shade of emphasis on the word, *square*. He could also say he is "on the LEVEL." These phrases are part of normal conversation, but with the right inflection, the other Mason understands.

The same lines could be used in bargaining for a deal on a car, or a home. Going to a jewelry store to buy gemstones, I would say to the manager,, "I hear I can get a square deal here." He (if he didn't know me to be a Mason personally) might say, "I see that you are a traveling man."

I would reply, "I am. I travel from west to east and from east to west again."

He might ask, "Why did you leave the west and travel to the east?"

I would respond, "In search of that which was lost."[22] That would do it. He would know I was a Master Mason, and I could usually get my gems for his cost! That may seem harmless in a minor business transaction, but imagine what would happen if this was done in a court of law! And it is being done, every day!

Another phrase (or gesture) which is only to be

Sign of a
Master Mason

Grand Hailing Sign
of Distress

used in extreme situations is the Grand Masonic Hailing Sign of Distress.

Our Masonic defendant in court might bury his head in his hands at some point and cry, "Oh Lord, my God, is there no help for the widow's son?" Although that might sound a bit odd to the average listener, it could be understood as a cry of anguish. If a Masonic judge or juror heard it, they would be honor bound (by a blood oath)[23] to acquit such a person, or at least fight for a hung jury!

The gesture which accompanies this (or it can be given alone, if necessary) is for the person to raise his arms over his head (almost as in a "hands up" position) and then lowering them in three stages, pivoting the forearms at the elbow until they are perpendicular to the floor, palms down.

Any Mason, seeing that gesture, (or hearing the above words) would be oath-bound to do anything possible to save the other Mason from danger, up to, but not including, the loss of his own life!

12

Lambskin Salvation

There are two "litmus tests" for checking the validity of any sect. The first is: How high a view of the Bible do they take? Is it the inerrant Word of God? Is every part of it perfect, all-sufficient for doctrinal understanding? (II Timothy 3:16)

The second is: What does the group teach about salvation? What must one do to be saved?

We have already given the Masons enough rope to hang themselves. They have a different god, and a different view of Jesus and His commandments. That should be enough to send any Christian fleeing from the Lodge like a scalded cat. However, to leave nothing to chance, we will briefly examine these two key issues.

That we should bring up the Bible may surprise many Masons. After all, is not the Bible one of "the three great lights of Masonry?" Are not the oaths sworn on a Bible? Is not the Bible opened with solemnity as part of every Lodge's opening ritual?

Are not passages from the Bible quoted copiously during ritual, and are not the very characters of Masonic drama taken from its pages?

DUST CATCHERS?

During the first week I came to work at our ministry, I was given a letter to answer from an irate Mason who had read our tracts. He expressed outrage at our statement that Masons don't believe in the Bible. He said, "The Bible is open at every Lodge meeting. I keep Bibles at home all the time!"

He kept Bibles? I had to smile sadly. It sounded like he was raising gerbils or something! From the rest of the fellow's letter, I could tell he knew very little about the Word of God. In fact, he knew less than the five and six year olds I had seen in our Sunday school class.

I could visualize a bookshelf or coffee table where a Bible lay. It was probably dusty, and only opened to record family milestones. It was probably a very nice Bible, but it must have creaked from lack of use. In short, it was very similar to the Bibles in all too many Christian homes.

This man seemed to have an almost superstitious regard for his Bible. It was like garlic being used to repel vampires. He assumed that as long as somehow, somewhere, a Bible was present, everything would be Christian.

The Bible on the altar of the Lodge I attended was never read. I, as the Senior Deacon (a Lodge officer), was forbidden to open it past Malachi. It had to stay in the Old Testament. So what does the Lodge really teach about the Bible? Is it the inerrant Word of God, or is it just a dust catcher?

THE AUTHORITIES SPEAK

Masons think that theirs is a Bible-based institution because Biblical themes are pervasive. It must be remembered, however, that the essence of a counterfeit is that it looks very much like the original. Therefore, we must examine closely the statements of Masonic ritual and authorities. Albert Pike, for example, wrote this about the Bible:

> The teachers, even of Christianity, are, in general, the most ignorant of the true meaning of that which they teach. There is no book of which so little is known as the Bible. To most who read it, it is as incomprehensible as the Sohar. (a Kabalistic book of Jewish mysticism—author)[1]

Elsewhere, Pike denigrates the Bible and waxes eloquent on the "Kabalah," a system of quasi-Jewish mysticism which is considered abhorrent to orthodox Jews. Kabalism is thought of as Jewish heresy because it denies monotheism (belief in one God) and teaches a Gnostic doctrine called "emanationism," the belief that an unknowable God manifests in ten emanations or ways.

Pike chooses to scuttle 4,000 years of Judeo-Christian orthodoxy:

> The Kabalah alone consecrates the alliance of the Universal Reason and the Divine Word...the Bible, with all the allegories it contains, expresses, in an incomplete and veiled manner only, the religious science of the Hebrews...

> The Hebrew books were written only to recall in memory the traditions, and they were written in symbols unintelligible to the profane. The Pentateuch and the prophetic

131

poems were merely elementary books of doctrine, morals or liturgy; and the true secret and traditional philosophy was only written afterwards, under veils still less transparent. Thus, was a second Bible born, unknown to, or rather uncomprehended by, the Christians; a collection, they say, of monstrous absurdities; a monument, the adept says, wherein is everything that the genius of philosophy and that of religion have ever formed or imagined of the sublime...One is filled with admiration, on penetrating into the Sanctuary of the Kabalah, at seeing a doctrine so logical, so simple, and at the same time so absolute.[2]

Amidst all that metaphysical prolixity is buried the idea that the Bible, as conceived by Christians, is relatively worthless compared to the deposit of hidden knowledge available to the initiate.

As a former Kabalistic "adept," I spent many years studying the Kabalah. When I finally let the Word of God speak to my life in plain English, it quickly tore down the metaphysical house of cards I had built for myself ! Obviously, Pike disdains the Bible.

Elsewhere, Pike calls the Church Fathers "dunces!"[3] Not a very promising beginning. Albert Mackey, 33° declares that:

The Bible is used among Freemasons as a symbol of the will of God, however it may be expressed. Therefore, whatever to any people expresses that will may be used as a substitute for the Bible in a Masonic Lodge. Thus, in a Lodge consisting entirely of Jews, the Old Testament alone may be placed upon the altar, and Turkish Freemasons make use of the Koran. Whether it be the Gospel to the Christians, the Pentateuch to the Israelite, the Koran to the Mussulman, or the Vedas to the Brahmin, it everywhere Masonically conveys the same idea—that of the symbolism of the Divine Will revealed to man.[4]

Turning to the most highly esteemed Henry W. Coil, we read again that the Bible is just a symbol of the "Volume of Sacred Law." Coil cautions:

The prevailing Masonic opinion is that the Bible is only a symbol of Divine Will, Law or Revelation, and not that its contents are Divine Law, inspired or revealed. So far, no responsible authority has held that a Freemason must believe the Bible or any part of it.[5]

THE MASONIC BIBLE SPEAKS

Right in the Masonic Bible, which is bestowed upon every Masonic initiate, there is an article by the Rev. Joseph Fort Newton entitled, *The Bible in Masonry*. In this article, which must be pretty authoritative to be given to each and every Mason at so solemn a moment, we find the following:

...like everything else in Masonry, the Bible, so rich in symbolism, is itself a symbol—that is, a part taken for the whole. It is a sovereign symbol of the Book of Faith, the Will of God as man has learned it in the midst of the years—that perpetual revelation of Himself which God is making mankind in every land and every age. Thus, by the very honor which Masonry pays to the Bible, IT TEACHES US TO REVERE EVERY BOOK OF FAITH in which man can find help for today and hope for the morrow, joining hands with the man of Islam as he takes oath on the Koran, and with the Hindu as he makes covenant with God upon the book that he loves best.

For Masonry knows, what so many forget, that religions are many, BUT RELIGION IS ONE...Therefore, it invites to its altar men of all faiths, knowing that, if they use DIFFERENT NAMES FOR "THE NAMELESS ONE OF A HUNDRED NAMES" they are yet praying to THE ONE GOD AND FATHER OF ALL; knowing also, that while they read different volumes, they are in fact reading the SAME VAST BOOK OF FAITH of man as revealed in the struggle and sorrow of the race in its quest for God. So that, great and noble as the Bible is, Masonry sees it as a symbol of that eternal Book of the Will of God...

Oliver Day Street, another scholar of the Lodge, tells us this:

No lodge with us should be opened without its (the Bible's) presence. Still, it is but a symbol; it represents divine truth in every form...

But the shadow must not be mistaken for the substance. There is nothing sacred or holy in the mere book. It is only ordinary paper...It is what it typifies that renders it sacred to us. Any other book having the same signification would do just as well....In fact, that book should be used which to the individual in question most fully represents divine truth.[6]

Later, the same author asserts quite categorically:

We dare assert that neither the Constitution,

Regulations, nor Ritual of any Grand Lodge in the world requires a belief in the teachings of the Bible...(we must) frankly acknowledge the Bible to be a symbol only. Those Christian Masons who would enforce belief in the teachings of the Bible have simply mistaken the symbol for the thing itself.[7]

Where does that leave the sincere, Bible-believing Christian? Evidently, "mistaken." How could a Christian align himself to such teaching?

This doctrine is obviously worlds apart from the Bible. Just as Masonry has a generic god, so it also has a generic "bible" of sorts. Following this logic, a Lodge composed entirely of Satanists would properly convene with a copy of *The Satanic Bible* on the altar. Mormons would convene with a *Book of Mormon* and so on. Any old "sacred book" would do. This makes utter hash out of the inerrancy of the scripture.

Freemasonry's devotion to the Bible is superficial at best; arrogant at the worst. Such an institution cannot be of God.

"WHAT MUST I DO TO BE SAVED?"

That one question, so vital and essential, has reverberated down through the ages. The Bible's answer is simple: "Believe on the Lord Jesus Christ and thou shalt be saved..." (Acts 16:31).

Freemasonry's answer, perhaps predictably, is not so simple. Since Masonry will not honor Jesus as Lord, it obviously cannot point to Him as the only way to salvation. Like every other cult, Masonry offers salvation by works.

Many Christian Masons are astounded to hear

that charge, as they are convinced that the Lodge nestles comfortably in the bosom of Protestantism. To those who need convincing, let us examine first the most authoritative level of Masonry, the Ritual Monitor. In the third degree lectures, we read the following instruction:

> ...that All-Seeing Eye, whom the sun, moon and stars obey, and under whose watchful care even comets perform their stupendous revolutions, beholds the inmost recesses of the human heart, and will reward us according to our works.[8]

That is quite plain, and totally contrary to the Bible which teaches that:

> For by grace are ye saved through faith, and that not of yourselves, it is the gift of God: *Not of works,* lest any man should boast.
>
> Ephesians 2:8-9

This means that the system of salvation operating in Freemasonry is not Biblical! In the same lecture, the Masonic symbol of the "Sword pointing to the human heart" is said to remind the Mason that "God will reward us according to what we do in this life."[9]

The culmination of third degree is a charge which is read to the new Master Mason and the Lodge. In part, they are exhorted:

> Thus, when dissolution draws nigh, and the cold winds of death come sighing around us...with joy shall we obey the summons of the Grand Warden of Heaven, and go from our labors on earth to...the Paradise of God. Then, by the benefit of the pass—a pure and blameless life—with a firm reliance on Divine Providence, shall we gain ready

admission into that Celestial Lodge above, where the Supreme Grand Warden forever presides...When, placed at his right hand, he will be pleased to pronounce us just and upright Masons...[10]

You see? Nowhere is Jesus mentioned, or the fact that no one is righteous. It is all salvation by works!

TRUSTING IN THE LAMBSKIN?

Though there is no reference to Jesus Christ, the Lamb of God, in their rites, Masons are exhorted to trust in something besides their good deeds. This is their Lambskin or white leather apron..."an emblem of innocence and the badge of a Mason."[11] This is the apron seen often when Masons appear in public in their regalia. It is given in the first degree and essentially remains with them throughout all rituals.

Listen to what the EA ritual says about this strange little apron:

In all ages the lamb has been deemed an emblem of innocence; he, therefore, who wears the Lambskin as a badge of Masonry is continually reminded of that purity of life and conduct which is necessary to obtain admittance into the Celestial Lodge above, where the Supreme Architect of the Universe presides.[12]

This teaches that we gain admission to heaven through works. It is not Biblical salvation, but "another gospel" (Galatians 1:8)! Elsewhere, the third degree ritual continues to harp on this lambskin apron.

...May the record of your whole life and actions be as pure and spotless as the fair emblem I have placed in your hands tonight.

137

> And when at that last great day your poor trembling soul stands naked and alone before the great white throne, may it be your portion to hear from Him who sitteth as the Judge Supreme the welcome words, "Well done, good and faithful servant, enter thou into the joy of thy Lord."

The Mason who may have turned his back on the Lamb of God is asked to trust his good works, symbolized by a lambskin!

There is a new key element in this ritual presentation. We still have salvation by "purity of life and rectitude of conduct," though God's Word says "There is none righteous, no, not one" (Romans 3:10). The Mason is herein promised that he will appear before the Great White Throne Judgement mentioned in the Book of Revelation.

The only problem with that teaching (and here we see the vile subtlety of Satan) is that, according to Revelation 20:11-12, the Great White Throne is the place of the judgement of *the damned!*

> And I saw a great white throne, and him that sat on it, from whose face the earth and the heaven fled away; and there was found no place for them. And I saw the dead, small and great, stand before God; and the books were opened: and another book was opened, which is the book of life: and the dead were judged out of those things which were written in the books, according to their works. And the sea gave up the dead which were in it; and death and hell delivered up the dead which were in them: and they were judged every man according to their works. And death and hell were cast into the lake of

138

> fire. This is the second death. And whosoever was not found written in the book of life was cast into the lake of fire.

The White Throne judgement is where men are judged by their works and then cast into the lake of fire, *because their works cannot save them.* The clues are in the ritual for anyone to see. If they knew the Bible, they would see that all Masonry is promising them is the lake of fire!

It should be evident to anyone who takes the Bible seriously that Masonry is another system of salvation. It does not esteem the Bible as the inerrant Word of God. It does not confess Jesus as Lord! It does not even worship the God of the Bible. As such, it comes under the curse which Paul mentioned in Galatians:

> I marvel that ye are so soon removed from him that called you into the grace of Christ unto another gospel: Which is not another; but there be some that trouble you, and would pervert the gospel of Christ. But though we, or an angel from heaven, preach any other gospel unto you than that which we have preached unto you, let him be accursed. As we said before, so say I now again, if any man preach any other gospel unto you than that ye have received, let him be accursed.   Galatians 1:6-9

This means that all Worshipful Masters and other Masonic instructors are under the curse of God, just like well known cultists such as the Mormons. They all preach salvation by works.

The difference is that Freemasonry is not recognized as a cult. Former Worshipful Master, (now Christian author) Jack Harris calls it, "The Invisible

Cult in our Midst." We may hear sermons on other cults, but few pastors or evangelists are willing to call Masonry a sin.

Thus far, we have examined the surface of Masonry and let its rites and authors speak for themselves. However, as with any "hidden thing of darkness," there is a lot more beneath the surface. Nothing this well crafted could have just sprung up by happenstance. Freemasonry may well be Satan's counterfeit, par excellence!

What are the origins of Masonry, and how did it come to be so evil? What is it about the Lodge that makes otherwise strong pastors tremble? How has it managed to capture some of the oldest denominations in Protestantism? In many cases, those churches have been transformed by Freemasonry into ecclesiastical toothless tigers fleeing from Satan on the spiritual battlefield.

In the second part of this book, we will go beneath the surface and explore the underbelly of Freemasonry. As cursed as its surface is, the concealed parts are even more abominable. We shall see how "the Brotherhood," as it is called, has roots sunk deep into the blackest pits of hell, Satanism and Witchcraft.

PART 2

The History of Masonry

The Structure of American Freemasonry
33° Grand Sovereign Inspector General

Order of Knights Templar

Order of the Knights of Malta

Order of the Red Cross

Super Excellent Master

Select Master

Royal Master

Royal Arch Mason

Most Excellent Master

Auxiliary Bodies
- The Shrine

(for women)
- Order of the Eastern Star
- Daughters of the Nile
- White Shrine

(for youth)
- De Molay
- Job's Daughters
- Rainbow Girls

Past Master

Mark Master

YORK RITE

32° Sublime Prince of the Royal Secret
31° Inspector Inquisitor
30° Knight Kadosh
29° Grand Scottish Knight of St. Andrew
28° Knight of the Sun, Adept
27° Knight Commander of the Temple
26° Prince of Mercy
25° Knight of the Brazen Serpent
24° Prince of the Tabernacle
23° Chief of the Tabernacle
22° Knight of the Royal Axe
21° Noachite or Prussian Knight
20° Master, Symbolic Lodge
19° Pontiff
18° Knight Rose Croix
17° Knight of the East & West
16° Prince of Jerusalem
15° Knight of the Sword
14° Perfect Elu
13° Royal Arch of Solomon
12° Master Architect
11° Elu of Twelve
10° Elu of Fifteen
 9° Elu of Nine
 8° Intendent of Building
 7° Provost and Judge
 6° Confidential Secretary
 5° Perfect Master
 4° Secret Master

SCOTTISH RITE

BLUE LODGE

Master Mason 3°
Fellowcraft 2°
Entered Apprentice 1°

13

Can an Evil Tree Bring Forth Good Fruit?

Masonry is worlds apart from true Christianity, and in spite of its good, its spiritual inheritance is a cursed thing. Good deeds do not make up for disobedience to God (I Samuel 15:22).

By way of warning, it is my intention to establish what I have learned from years of experience in both the occult and Freemasonry: that the Lodge is part of a huge network of world-wide Witchcraft organizations.

It could be said, with poetic license, that Freemasonry is the world's largest coven of witches. This statement would doubtless astound most Masons. However, appearances can be deceiving, and Masonry, like any secret society, has layers within layers—it is a Chinese puzzle box of evil.

Each box unveils a stronger increment of iniquity. Yet these "boxes" remain shut to Masons who do not advance to higher levels. Since most Masons do not rise higher than the 32° in the Scottish Rite or the Knight Templar in the York Rite, they get no more than glimpses behind the veil.

Those glimpses should be enough to cause any Christian to flee from the Lodge. These Masons are being deliberately deceived by their leaders. Albert Pike remarks:

> The Blue Degrees are but the outer court or portico of the Temple. Part of the symbols are displayed there to the initiate, but he is intentionally misled by false interpretations. It is not intended that he shall understand them, but it is intended that he shall imagine he understands them.[1]

The Blue Lodge is the "outer court." This shows how deceptive Masonry is. The third degree candidate is informed that the Lodge of Master Masons meets in a place representing "...the unfinished Sanctum Sanctorum or Holy of Holies of King Solomon's temple."[2]

The impression given was that the Lodge represented the inner sanctum of the temple. Now Pike tells us that we are, at that point, only at the outer portal. This is typical of secret societies. Just when you think you have been told everything, a new box opens, and a new level of secrets awaits you. It is the old "bait and switch" again.

HIDING THE REAL TRUTHS?

The odd "secrets" of low level Masonry (handshakes and words) are just red herrings, designed to make the "rubes" happy. They are not the real secrets of Masonry! Pike teaches that:

> Masonry, like all the religions, all the
> Mysteries, Hermeticism and Alchemy,
> conceals its secrets from all except the
> Adepts and Sages, or the Elect, and uses false
> explanations and misinterpretations of its
> symbols to mislead those who deserve only
> to be misled; to conceal the Truth, which it
> calls Light, from them, and to draw them
> away from it. Truth is not for those who are
> unworthy or unable to receive it, or would
> pervert it.[3]

With statements like that on the record for
anyone to read, how can the honest Mason know
what to believe? All sorts of benign meanings are
assigned to the working tools of the Mason in the
lower degrees. For example, he is told in first degree
that the meaning of the square and compasses is:

> ...the square, to square our actions; the
> compasses, to circumscribe and keep us
> within bounds of all mankind, but more
> especially with a brother Mason.[4]

However, should the candidate take the time to
read some of the books in his Lodge library, he
would find more disturbing meanings. Here the
square and compasses are related to the "point
within the circle." The deeper level of symbolism is
revealed by Albert Mackey, 33°:

> The point within the circle is an...important
> symbol in Freemasonry, but it has been
> debased in the interpretation of it in the
> modern lectures (given in lodges) that the
> sooner that interpretation is forgotten by the
> Masonic student, the better will it be. The
> symbol is really a beautiful but somewhat
> abstruse allusion to the old sun-worship, and
> introduces us for the first time to that modifi-

cation of it, known among the ancients as the worship of the phallus.[5]

In his definitive encyclopedia, the same author writes:

The phallus was a sculptured representation of the male organ of generation and the worship of it is said to have originated in Egypt. In the Mysteries...we find the remote origin of the point within a circle, an ancient symbol which was first adopted by the old sun worshippers...and incorporated into the symbolism of Freemasonry.[6]

Now we see that the central symbols of Freemasonry actually represent the human reproductive organs! In what is called esoteric Freemasonry, we learn that the square is the symbol of the lingam (or god force in Witchcraft), and the compasses are the symbol of the female organs, called the yoni or shakti (goddess force) by occultists.[7]

So here's the Christian Mason, attending Lodge and worshiping at an altar with symbols of the male and female organs entwined over an open Bible! What blasphemy! Talk about being destroyed for lack of knowledge (Hosea 4:6)!

FOUL ROOTS?

Let us see the actual soil from which Masonry springs, from the words of its ritual and its teachers. According to the third degree, the first Freemason was Tubal Cain.[8] In the Bible, we find that Tubal Cain is descended from the accursed line of Cain (Genesis 4:17-22).

Interestingly, in light of the Masonic obsession with murder, we find that Tubal Cain's father, Lamech, is the first person to boast about murder

146

(Genesis 4:23-24). What a nice home for the founder of Freemasonry to be raised in!

The next Masonic "saint" spoken of is Nimrod, who is described by Mackey as "one of the founders of Masonry."[9] In Genesis 10:8-9, Nimrod is called a "mighty one in the earth" and a "mighty hunter before the Lord." Expositors generally identify Nimrod as the founder of Babylon, and architect of the Tower of Babel.[10] That certainly qualifies him as a Mason, but it doesn't do much for his reputation.

The builders of the Tower of Babel, in Genesis 11:4-5, 7-8 present a proud and astonishingly modern Masonic mentality:

> Go to, let us build us a city and a tower, whose top may reach unto heaven; and let us make us a name, lest we be scattered abroad upon the face of the whole earth.

> And the LORD came down to see the city and the tower, which the children of men builded.

> Go to, let us go down, and there confound their language, that they may not understand one another's speech.

> So the LORD scattered them abroad from thence upon the face of all the earth: and they left off to build the city.

It is obvious the Lord did not approve of Nimrod's plans. This was the first attempt at what today we call globalism, an attempt to unify all people under one common religio-political system. In the Bible, Babylon stood for evil against Jerusalem, the city of God.

Nimrod helped found the fountainhead of all cults.[11] The religion which sprang up around him

147

and his queen, Semiramis, became the prototype for virtually all cults.

The rites of the "slain and risen god or rite of the divine king" form one of the core rituals of Witchcraft and all fertility cults.[12] Nimrod was slain and cut into pieces by his grandfather, Ham. He and Semiramis became the first of a long line of mother-goddess and father-god pairs. Their son, little Nimrod, Jr. was the supposed resurrected father-god.[13]

We see various trinities down through the centuries: Isis, Osiris and Horus, and sadly enough the cult of the Madonna and child from Catholicism. The idea of a slain god who is raised from the dead is one of the central themes of all mystery cults. Amazingly, it is the ritual core of Freemasonry as well.

Nimrod is probably one of the two or three most evil men in the entire Old Testament, and to identify him as a Masonic leader does little to enhance the spiritual wholesomeness of the Lodge.

THE TEMPLE OF SOLOMON AND THE "WIDOW'S SON"

The next significant milestone in Masonic history is the time of the building of King Solomon's temple. This event provides the backdrop for a large percentage of the ritual in the Lodge. However, there is no real indication that the Masonry mentioned here actually existed in the days of Solomon. In fact, most serious Masonic historians discount as myths the legends that Solomon was a Master Mason.

Nevertheless, Masons place incredible store in

the characters which make up the ritual. They are: King Solomon, Hiram, King of Tyre, and Hiram Abif, the Widow's Son. All three are mentioned in the Bible, although Hiram Abif is only a footnote in the building of Solomon's temple.

Hiram Abif was supposedly in possession of the "Master's Word." Although the Bible only identifies him as a craftsman who did the metal work in the temple, Masonic lore makes him seem like the architect of the entire project.

To make a long ritual very short, Hiram is accosted by three "ruffians," Jubelo, Jubelah and Jubelum, each of whom are Fellowcraft (i.e., second degree Masons) who want the secret of the Master's Word because the temple is near completion.

The first ruffian smites him on the throat, and Hiram staggers around the perimeter of the temple, only to be attacked by the second ruffian who strikes him in the chest. He reels a bit further and then is brained with a setting maul by the third ruffian and falls dead.

The three thugs bury the body in the rubble of the temple and later, lug it out of town and bury it on a hill under an acacia tree. Solomon and King Hiram send out a search party and eventually, the body is found on the hill of Mt. Moriah. They both go there.

After a lot of ritualized fussing around, Solomon takes the decomposing right hand of Hiram Abif by the "Strong Grip of the Lion's Paw," the Master Mason grip, and hauls his carcass out of the ground, apparently resurrecting him (although this is never clear).[14] This act is the central ritual of all Freemasonry. The trouble is, it never happened.

All Masonry moves onward from the resurrection of Hiram. Supposedly, the secrets of Masonry have been kept until they were codified by what is called the "Mother Lodge" in England in 1717. This Mother Lodge is the source of most American Masonic orders today, although there are other forms of Freemasonry.

This is the foundation of Freemasonry—the history it chooses to write for itself. As you can see, it is built upon either vile characters from the Bible, or upon a tissue of myth woven around certain noble Bible heroes. It is not a very auspicious beginning.

14

No Help for the Widow's Son?

Masonry's attempts to anchor itself into Biblical history and doctrine run aground on the rocks of reality. But why should we be surprised?

We already know that the material taught in the Blue Lodge degrees is designed to deceive the Mason. Only those Masons who pursue the higher degrees and are willing to go into musty old libraries (as I did) see the real source of Masonry, both doctrinally and historically!

I remember the first time I went into the huge Scottish Rite cathedral library on Van Buren St. in Milwaukee. It was a sunny afternoon and the beautifully appointed place gleamed with tones of old wood and rich carpeting. It also had an aroma I had grown to love—the smell of old books!

There I saw my first copy of Pike's *Morals and Dogma,* and Mackey's works. I also found books by Manly P. Hall and other occultists. There was even an expensive edition of Hall's monumental work,

The Secret Teachings of All Ages and Countries. Vast in its scope, this book deals with every conceivable form of mysticism, sorcery and idolatry under the sun!

I could tell these books had not been read much. In fact, I was the first to take out Pike's book in months. Nevertheless, Morals and Dogma was the "bible" of the Scottish Rite. It was in those books, among others, that I found confirmation of what I had been told for years by my teachers in Witchcraft. I saw clearly that Masonry was a form of the "Old Religion" of devil worship.

That is a strong accusation, but the books are there for any Mason to check out. They are probably gathering dust in scores of libraries just like the one in Milwaukee! Many Blue Lodges have good libraries as well, depending on their size. The question is, does the average Mason have the courage to dig into these books and find out for himself where his Lodge comes from?

To aid in that quest, we will examine some of the concepts which make Freemasonry a form of Witchcraft, and look at the teachings of Masonry's leaders on the subject.

A MYSTERY RELIGION

Freemasonry has been vaunted by its authorities to be a form of the "Ancient Mysteries."[1] That sounds fascinating, but what does it mean? On its most superficial level, a mystery religion is a religion in which secret doctrines are kept from the public.

Christianity is not a mystery religion. All its elements are readily available to non-Christians. It is an "open book." Its rites may be attended by anyone.

In mystery religions, all or part of the religion are zealously concealed from the "profane" (i.e., outsiders). Two examples of mystery religions in America today are the Mormon church and the Freemasons, although most Americans would not recognize them as such.

In Mormonism, the temple rituals are available to only a select few. No outsiders (or "unworthy" Mormons) may enter temples after they are dedicated, although most Mormon worship is open to the public.

Freemasonry is a different matter. Except for public appearances (i.e., parades, funerals) and ceremonies like cornerstone layings, all Masonry is closed to the profane. This makes it a consummate mystery religion. But what are "the mysteries"? The dictionary defines them as:

> Ancient religions that admitted candidates by secret rites and rituals the meaning of which was known only to initiated worshippers.[2]

Virtually all ancient cultures, whether Roman, Celtic or Egyptian, had some sort of mystery religion.[3] Although these groups were called by different names in different parts of the world, they all had certain elements in common.

TWO WORLD VIEWS

The basic features of this Pagan Mystery Religion (Revelation 17:5) are:

1) Polytheism (a belief in many gods or goddesses) or pantheism (a belief that God IS the universe, and that is all He is).

2) A cyclical view of history (the belief that there are eternal, repeatable cycles of life).

3) The veneration or worship of the regenerative processes of nature (sex) as the "sacred mystery."

This contrasts with Biblical Christianity, which holds to:

1) Monotheism (belief in one God).

2) A linear view of history (the belief that time has a beginning and an end, and that God has moved, is moving, and will move into history in spectacular, miraculous invasions of His power and love).

3) The worship of God through His Son Jesus Christ.

Obviously, these two philosophical approaches are quite different, which is another reason why a Christian cannot be a Mason and be true to both sets of beliefs.

What is meant by the second and third constituent of the mystery religions? The cyclical view of history sees time as an eternal round of seasons, and the seasons themselves as reflections of divine interactions in the cosmos. Thus, time will never really end. It is sort of a spiral upwards through cycles of reincarnation.

Though all participants in mystery cults do not necessarily believe in reincarnation, Masonic scholars find it an essential part of their belief system.[4] The seasons are seen as a pageant of the gods.

Perhaps the best known form of this pageant is in the Eleusinian mysteries, where the cycle of the corn goddess, Demeter, and her daughter, Persephone, are felt to be manifestations of the waxing and waning of the fertility of the earth.[5] Persephone is stolen by the dark lord of the underworld and taken there. Demeter strikes a bargain with him and is allowed to have her daughter back for the spring.

This is the explanation for seasonal change. The goddess mourns for the loss of Persephone during the fall and winter by producing—you guessed it! Fall and winter! Contained within this myth is another, darker myth. It is this myth which is more central to Freemasonry.

THE WORSHIP OF SEXUALITY!

Tied up intricately with the pageant of the seasons is the concern for the fertility of the land. Today most of us get our food from the supermarket, but for our ancestors, famine and drought were very real dangers. The coming of the spring was a time of rejoicing as buds started appearing. Even in our urbanized culture, who doesn't rejoice at seeing the first apple blossoms after winter?

There is nothing wrong with enjoying the spring-time, or celebrating the beauties that God has given us. It is all to be received with thanksgiving (I Timothy 4:3). However, the difference between the Judeo-Christian world view and that of the mystery religions is that the "mysteries" believed that the god resided right in the processes of fertility. They believed that the elements of plant, animal and human reproduction were godlike!

The great mystery of the mystery religions was the worship of sex! Part of this mystery was involved in the mechanics of human reproduction itself. There is a doctrine in Witchcraft called the Hermetic Maxim: "As above, so below, but after a different manner." This idea is the basis of much of magic and astrology.[6] This means that human reproduction is a reflection of the larger universe.

The people who put together the mysteries noted the differences between male and female sexuality. Using this Hermetic maxim, they deduced an eternal goddess and a finite god who died and needed to be resurrected yearly, and tied this into the seasons.

They made the sun (which seems to move south and weaken in power in winter) a symbol of the god. They made the moon (which waxes and wanes with a woman's cycle) the symbol of the goddess' daughter; and the earth, with its obvious stolidity and fertility, the symbol of the eternal mother goddess.

THE "WIDOW'S SON"—A DYING GOD?

In the thinking of these people, the god was "born" every year at the winter solstice (December 23, when the sun is farthest away). The sun god "impregnates" the earth on the summer solstice (June 22, when the sun is most powerful) and then begins to wane and die.

Many European pagans believe the god dies around the time of the feast of Samhain or Halloween (October 31). Then the goddess gives birth to a new "baby god" on December 23 and the cycle begins anew.

This is the "bare bones" of the cycle, but there are embellishments which are relevant to Masonry. In a well known form of this myth, we have the story of Isis and Osiris. Isis is the key goddess of Egypt. Isis was the sister, lover, wife and possibly even the mother of Osiris.

Osiris had a wicked brother, Set, who murdered him.[7] Isis searched for him, weeping, as the land of Egypt dried up. Eventually, Isis found her lost lover, and breathed life back into him.

Set, however, was not a happy camper! He hacked up Osiris into pieces and scattered them to the four winds. The legend recounts that Isis was grief-stricken and tracked down all the parts. However, she could not find one vital part (knowing the pagan mind, guess which part!).[8] Because of this lacking part, she was not able to restore him to life.

Instead, she enthroned him in the underworld as the Lord of the Dead, while Set became the Egyptian version of the devil. However, Isis and Osiris had a son, Horus, who eventually slays Set and avenges his father.[9]

In this legend, we see the close correlations between the Masonic legend of Hiram Abif and the legend of Osiris.

Both are "Widow's sons" (Isis being both widow and mother of Osiris, in her form of Nephthys). Both are craftsmen.[10] Both are slain by wicked men, who are, in turn, slain in vengeance. Both have unsuccessful resurrection attempts made upon them. In the deaths of both, a thing of great power is lost (the Master's Word and Osiris' organ of generation[11]).

Both are lamented with great ceremony. In the Lodge rite, King Solomon makes the "Grand Hailing Sign of Distress," and says, "O Lord, my God, I fear the Master's Word is forever lost."[12] Both are buried, dug up, buried again, and finally buried a third time, after an attempt to raise them from the dead.[13]

However, because Masonry is a masculine form of the mysteries, the goddess' role is downplayed. Nevertheless, it is still contained in some of the symbolism of the "Hiramic legend."

The third degree lecture discusses the monument erected to the memory of Hiram (never mind that no Hebrew would build a monument of statues).

> ...Masonic tradition informs us that there was erected to his memory a Masonic monument, consisting of "a beautiful virgin, weeping over a broken column, before her, a book was open; in her right hand a sprig of acacia, in her left an urn; behind her stands Time, unfolding and counting the ringlets of her hair."[13]

Isis was both virgin and mother, so the "beautiful virgin" is Isis weeping. The broken column is the missing member of Osiris, the acacia, an allusion to the eternal life preached by the Egyptians as well as the fertility cults' emphasis on vegetation. The urn is an evocation of the canopic jars[14] used in Egyptian funerals to store the vital organs of mummies.

Finally, we have "Time," the god Saturn, a later form of the mysterious and evil god, Set. In astrology, Saturn is called the "greater malefic," or greater evil. A weirdly disturbing set of coincidences, eh? There are darker aspects yet to this "mysterious myth."

THE EXPERTS SPEAK AGAIN

Some people might find these connections tenuous, and feel we are exaggerating the connections between Freemasonry and these cults.

However, Masonic experts agree with this analysis. Pike teaches the following:

> In the Mysteries was also taught the division of the Universal Cause into an Active and a Passive cause; of which two, Osiris and Isis,—the heavens and the earth were symbols...the Active and Passive Principles of the Universe were commonly symbolized by the generative parts of man and woman; to which, in remote ages, no idea of indecency was attached; the Phallus and Cteis, emblems of generation and production, and which, as such appeared in the Mysteries...

> ...which were symbolized by what we now term Gemini, the Twins, at the remote period when the Sun was in that Sign in the Vernal Equinox, and when they were Male and Female; and of which the Phallus was perhaps taken from the generative organs of the Bull.[15]

I apologize for the content of this material, but it serves to contrast the attitude of this "Plato of Freemasonry," towards these pornographic cults and then compare it to his attitude towards those "dunces" (his words), the Christian church Fathers. One paragraph later, he writes:

> The Christian Fathers contented themselves with reviling and ridiculing the use of these emblems (sexual symbols). But as they in the earlier times created no indecent ideas, and were worn alike by the most innocent youths and virtuous women, it will be far wiser for us to seek to penetrate their meaning.[16]

Elsewhere, Pike proclaims:

> Masonry, successor of the Mysteries, still follows the ancient manner of teaching. Her ceremonies are like the ancient mystic shows.[17]

Albert Mackey also rhapsodizes about these ancient Mysteries:

> ...the secret worship rites of the Pagan gods. Each of the pagan gods had, besides the public and open, a secret worship paid to him, to which none were admitted but those who had been selected by preparatory ceremonies called Initiation.[18]

Finally, Pike makes the following claim:

> The Occult Science of the Ancient Magi was concealed under the shadows of the Ancient Mysteries; it was imperfectly revealed or rather disfigured by the Gnostics; it is guessed at under obscurities that cover the pretended crimes of the Templars; and it is found enveloped in enigmas that seem impenetrable in the Rites of the Highest Masonry.[19]

Freemasonry wishes to link itself with the ancient pagan Mysteries. We shall follow this suggested trail that Pike and others have left us, and find out how these bizarre and depraved rites have found their way into countless sedate Lodge rooms throughout Christendom.

15

The Children of Baphomet

The connection between Masonry and the Sirius cult is central to understanding the Lodge's peril. Most Masons have never heard of Sirius. They do not realize that when they enter the Lodge and kneel at its altars, they are submitting themselves and their families to the authority of the Dog-star god, Set!

This is not to be taken lightly! We need to see how Set got from the temples of Egypt to the modern Lodges. We will now briefly trace the historical and spiritual roots of Masonry.

Contrary to Masonic lore, there is no actual evidence for the Craft existing prior to the Middle Ages. There are obvious attempts on the part of Masonry to evoke the ancient mysteries. This fact, however, in no way weakens the spiritual peril of Freemasonry today!

It is like Europe following World War II. Folks dug up buried bombshells which had never exploded. If they left the bomb alone, all was well.

However, if it was moved, it could blow up in their faces, though it had lain in the earth for years!

Similarly, although there may be no temporal link between pre-Christian cults and Masonry, their use of these ancient idols is very much like tinkering with a long-buried bomb.

I recall a conversation I had with a fellow Mason who was an occultist. He had a huge collection of occult books, including the writings of Mason/magician Manly P. Hall. He was an esteemed Mason and was about to become Worshipful Master.

This man made the remark that most Masons he knew were like children who had found a chess set. They were fascinated by the elegant shapes of the pieces. Perhaps they began playing checkers with the chess pieces—unaware of their power. They went through the motions without comprehension.

The typical Mason is toying with idols of which he has no understanding. These idols are primed by the power of Satan to detonate in his face! The symbols themselves need have no intrinsic power, but the sin that is involved in their use will open the door to demonic access for the Mason and his family.

CLASSICAL ROOTS?

The lineage of Masonry can be traced to Nimrod. He was the fountainhead of all the Mystery religions involving the death and resurrection of a sexual god. However, according to Masonic scholars, the earliest (c.500 BC) emergence of a secret society of builders quite similar to the Masons would have been the "Dionysian Artificers."[1] (Artificer is another word for a builder.)

Dionysos was the horned Greek god of wine and madness, and was a slain and risen god conceived by a virgin, being impregnated by a god, Zeus.[2] The Dionysian rites had a lot in common with witches' sabbats[3] and with Masonry, including enactment of slaying and resurrection, oaths and threats of death.

Also worthy of mention is the Pythagorean school. The philosopher and mathematician, Pythagoras (582-507 BC), is referred to in the Masonic ritual as the Masons' "ancient friend and brother."[4] Pythagoras was a genius, but he was also an occultist, a product of pagan Greek culture.[5] He was the founder of a brotherhood that believed in Reincarnation.[6]

THE THRONE OF THE PEACOCK

Another source of Masonry can be traced to the Middle East near the site of Ninevah and modern Baghdad. A group called the Yetzidis worshipped "the Peacock Angel," and like the Masons, had layers of secrets within their faith. No one seems to know how old the Yetzidi cult is, but persecution by the Muslims began in the 13th century.

Their god's secret name was Shaitan, the Arabic word for Satan.[7] It is a variation of the Hebrew, ha satan, used in Job 1:6 for Satan. The peacock was Shaitan's symbol because of its pride.

Only fragments of their writings remain.[8] They seem to have been unabashed Satan worshipers, although there is no evidence that they performed human sacrifices, as did some of the Mystery cults. Some speculate that the Yetzidis may have been somehow involved with the next link in this bizarre chain—the cult of Hassan I Sabah.

163

THE "OLD MAN OF THE MOUNTAINS"

Hassan I Sabah was a most intriguing figure in history. Called "The Old Man of the Mountains," he was a leader in the Islamic heresy called Ishmaelianism. Islam was bursting with fanaticism, and to be a heretic was to court death by the sword.

Hassan was cunning and realized that he was outnumbered by the more "orthodox" Muslims. He knew the only way he could win was by subversion. He created an elite order of soldiers with drugs, especially Alamout Hashish, and what we would call today mind programming.

Such programmed men were willing to die for their new master. Hassan's Order was called the Hashishim, or Assassins. The word assassin is traced directly to him.[9] He was a malignant genius who seems to have invented many of our modern espionage techniques—notably the mole. He would "bury" his men in deep cover in enemy courts and years later, in response to a signal, the "mole" would bury the curved dagger of the Assassins in the target's heart.

When Hassan died, he left behind an extraordinary legacy, because there was cross-pollenization between his ideas and the ideas of his European counterpart, the Knights Templar.[10] Like disease-carrying rats, the Templars brought back to Europe Hassan's secrets where they spread like wildfire.

CHILDREN OF BAPHOMET

The Templars are a fascinating curiosity which have excited the interest of more people over the centuries than any other military order. Their actual history is controversial and their exploits ranged from the noble to the depraved. No history of the Freemasons can be understood apart from the Templars. Oddly enough, the Templars have also inspired the interest of an incredible host of occultists, witches, satanists and other curious folk.

The Templars began with a noble idea—to protect pilgrims on their way to visit the Holy Land during the Crusades. The "Poor Knights of the Temple of Solomon" (as they were known) began in 1118 as an order of warrior monks.[11] Their first Grand Master (note that title) was Hugues de Payens.

Unlike other knightly orders, the Templars were monks, and required vows. They represented "the first time in...history (that) soldiers would live as monks."[12] Part of the reason for their name was that they built their headquarters at the site believed to be the place of the Temple in Jerusalem, a key place in Masonic legend.

However, things went sour. The Order grew quite rich and corrupt. The Templars established themselves as the most powerful institution in Christendom—easily the mightiest institution in all of Europe, equalled, but not surpassed by the papacy.[13] Then, in 1291, disaster struck. The crusaders were routed and the Templars were driven out.

A ROYAL PLOT

The Templars were fighting men, with no battles to fight. They returned apparently vanquished from a long, unpopular war in an distant land. Their morale was not very high, but they returned to Europe with an incredible fortune and a system of garrisons, and the finest fleets in the world.

They aroused the wrath of Philippe IV of France, one of the most powerful kings in Europe. He had succeeded in murdering one pope, poisoning a second, and installing a hand-picked successor, Clement V. He then took the papacy, lock, stock and papal tiara out of Rome and plunked it down in Avignon, France.[14]

Philippe had feared the Templars but was determined to have their wealth. Confident that his "pet pope" would cooperate, the king arrested all French Templars.

On Friday, October 13, 1307, dawn raids began. All the property that belonged to the Templars was confiscated. Although most Templars were arrested, their wealth was not found. Apparently the Order knew the raid was coming, for the Grand Master, Jacques de Molay, had many of the order's books burned.[15] Horrid oaths of secrecy guarded the nature of the actual Templar rites.

This lack of source material has created the controversy. Under Philippe, many of the knights were tortured. Obviously, any confession extracted under torture must be regarded with suspicion. However, there are certain elements in common which lead one to think there may be a kernel of truth to some of the confessions.

THE TALKING HEAD

The Templars were accused of practicing black magic, pederasty, homosexuality, murdering babies, blaspheming the name of Jesus, and giving each other "obscene" kisses. [16]

Also frequently mentioned is the allegation that the Templars worshiped a mysterious idol called Baphomet. This idol was described in various ways: a man with the head of a goat, a head with three faces, or a head with a beard, which taught the knights of the Order magical secrets.

BAPHOMET

Let's look at this mysterious figure closely. A common theory for the origin of Baphomet is that it is a variation of the name, Muhammed (Mahomet).[17] Even if this is the least bothersome of the explanations, it is still of grave concern. Worshipping

Baphomet

Muhammed would be both political and spiritual treason.

A second explanation is that the name is a riddle, called Notariqon. It is a way the Kabalists (ceremonial magicians) wrote words backwards and as acronyms. In Notariqon, Baphomet is rendered "Tem.-O.-H.-P.-Ab." In Latin, it means "Templi Omnium Hominum Pacis Abbas, or Father of the Temple of the Peace of All Men."[17]

That doesn't sound bad, except in the context of the age in which it is found. For knights in the midst of a religious war to worship a strange talking head called the "Father of the Temple of Peace of all Men" does sound odd, especially when we look at Freemasonry, the New Age movement, and their doctrines of "the fatherhood of God and the brotherhood of men."

Leading Islamic mystic, Idris Shah, claims that Baphomet is derived from the Arabic Abufihamat, which means "Father of Understanding."[18] This fits quite well with the theory that Templars were a survival of a Mid-eastern Gnostic sect of phallic-worshippers which believed in salvation through gnosis (or knowledge).[19]

Another theory advanced is that the name is actually Bapho-Mitras, meaning "Father Mithras."[20] Mithras was a sun god worshipped in Rome about the time of the rise of Christianity. He was regarded as a rival to Jesus, and his birthday was December 25.[21] Mithras was represented as a man with the head of a bull or as a bull-slayer, close to the one version of the idol as a goat-headed man.

The Freemason, satanist and pervert Aleister

Crowley took the name Baphomet when he assumed leadership of the occult/Masonic organization, the O.T.O. (Order of Eastern Templars). Crowley wrote of Baphomet as representing a type of phallic god.[22]

Kenneth Grant, the modern world leader of the O.T.O., claims that Baphomet conceals a formula of homosexual sex magic from the Templars.[22] Another world-class occultist of the last generation, the late Dion Fortune said Baphomet stands for a practice, which in its debased form, was "one of the causes of the degeneration of the Greek Mysteries."[23]

A Masonic scholar, Manly P. Hall, clearly identifies Baphomet with the satanic "Goat of Mendes," probably the best known representation of Lucifer in all occultism. He says they probably obtained it from the Arabs.[24]

THE HORNED GOD

These are deep waters which reveal the multiple layers of meaning used in mystery cults. However, two common threads also emerge in these confessions—the disembodied male head, and the horned man-animal.

The charges drawn up by the Inquisition against the Order in 1308 said they worshipped a head which had the power to make the land fertile.[25] These qualities of the head fit with the ancient Druidic worship of the severed head, the head of Bran the Blessed, a god from the Welsh poem, the Mabinogion.[26] They fit right into the Celtic Witchcraft pantheon, especially in terms of their fertility bringing qualities.

Baphomet is another name for the god of the

Witches, which is the god of the Templars as well! This god is also known as the "Horned God" because he is often shown with horns, or in the appearance of a horned animal (stag, goat or bull).[27]

This god is worshipped today in covens all over the world and may be one of the oldest idols in history. Cave paintings in Ariége, France show shaman or "sorcerer" wearing a horned costume.[28] Even Nimrod is often shown wearing a horned headdress.[29]

Another very common form of the god of the Witches in the British Isles is called "The Green Man."[30] This figure often has a green head with vines issuing from its mouth—and it never has a body! The resemblances here are too striking to be entirely coincidence.

It is important to remember that the original priesthood of pagan Britain were the Druids, and they had what were called "bardic colleges" (sort of satanic seminaries) all over France, Brittany and the Mediterranean almost a thousand years before the Templars came along. As we shall see, the Templar influence also ran deep in the British Isles.

DENYING JESUS CHRIST

One of the more serious, yet incredible charges flung at the Templars was that they denied Jesus as part of their initiation rituals. Since the Order was supposedly Christian, sworn to defend the holy places in Jerusalem and protect pilgrims, how could they be involved in something blasphemous? Strikingly, a key confession of this sort of denial was not extracted under torture, but in the slightly more "civilized" venue of England.

We may never know the full extent of the truth or falsehood of the charges against the Templars. Many of the knights were burned at the stake rather than confess these things.

In 1314, the Grand Master Jacques de Molay met the same fate, denying all charges. Legend has it that as he was being executed, he maintained his innocence and proclaimed that both the king, Philippe, and the pope would be dead within the year! The curse worked, for both the king and the pope did die within the year.

Whether the curse was simply a cry for divine justice or a real witchcraft curse is an unanswered question.

Nor can the deaths of the two potentates be ascribed to supernatural means with any assurance. If the Templars had an alliance with the Hashishim, they may have picked up some assassination techniques along the way. It is therefore possible that both the king and the pope were assassinated to continue the "legend" of the Templar's formidable power, even from beyond the grave.

If that was the intent, it has succeeded beyond anyone's wildest dreams. Templary has been one of the most enduring themes in secret societies and occult orders, up to the present day.

There is still one more ingredient to be added to the mixture before we are ready to have full-blown Grand Lodge Masonry as we know it today. We shall look at it next.

16

The Brethren of the Rosey Cross

Masonic historians admit a connection with the Rosicrucians. Nearly all agree that Freemasonry owes much to certain occult societies or groups that flourished (often in secret) during the late Middle Ages. Chief among these were the Rosicrucians and the Knights Templar.[1]

Some may think the connections between the Templars and Freemasonry are a bit farfetched, but the links between the two secret societies are more than conjecture. They are 800 hundred years old.

For example, the old Templar castle at Athlit was excavated in the Holy Land. The castle was built in 1218 and abandoned in 1291. Some of the tombstones contained markings—a mason's square, a plumb stone and a setting maul. With only two known exceptions, this is the earliest available evidence of graves bearing Masonic symbols![2]

No institution as powerful as the Templars could just evaporate. Some authors speculate that some of the Templars ended up in Scotland, where their traditions became an integral part of the development of so-called "Scottish Rite" Masonry.[3]

Imitators of the Templars arose quickly. The most notable of them was the Order of the Garter.[4] In 1348, just a few years after the fall of the Templars, Edward III of England created this order, which still exists today.

There is a legend behind the founding of the Order of the Garter by Edward. Supposedly, the king had been dancing with a lady in the presence of his court. Suddenly the lady's garter dropped to the floor. The incident shocked the court, and all the dancing ceased. Edward gallantly knelt and placed the garter on his own leg saying, "Honi Soit qui mal y pense," which means, "Shame be to him who thinks evil of it."[5]

In honor of the occasion, the king started the Order of the Garter, and the phrase he spoke became its motto. He created the Order with 26 knights (13 times 2). This rather strange event becomes even stranger when one realizes that in the 14th century people were not shocked by ladies' undergarments. This was boisterous semi-pagan England, and that garter caused shock for an entirely different reason.

The garter was (and still is) the symbol of a witch high priestess.[6] When a high priestess becomes a "witch queen," that is, when her coven of witches splits off a daughter coven with its own priestess, she acquires a silver, crescent moon-shaped buckle on her garter. For each coven created after that, a new buckle is added. Some historians speculate that the lady with whom Edward was dancing was a witch queen.[7] Dropping her garter in a court that was nominally Christian could have been an occasion for the woman to be arrested.

By expressing approval, Edward gave the lady and her religion his blessing. He may have meant, "Shame be to him who thinks evil of witchcraft," which is borne out by the king's choice of two groups of 13 knights, 13 being both the size of a coven and the number of moon feasts (esbats) in a given year.

To this day, the monarch of England is the head of the Order of the Garter, (as well as a patron of Masonry) and when invested with all the regalia of the Order wears a mantle with 168 garters on it, plus one actually worn on his or her leg. This equals 169—thirteen times thirteen. The queen of England may knowingly or unknowingly be the witch queen of her nation!

What happened to the inner teachings of the Templar Order in the midst of all these copies? Where did this strange blend of Christianity, the Hashishim and the Yetzidis end up?

THE FAMA FRATERNITATIS

Tracking a thread throughout history which is woven into the warp and woof of Masonry, its strands have two common ideas:

> 1) Incorporation of elements from the mystery religions, especially the idea of a hero-god who is slain and resurrected by occult power, and

2) The idea of a hidden fraternity with secrets from antiquity. Membership in this fraternity was obtainable only through a perilous initiation.

In terms of both history and occult lore, the clearest immediate successor to the Templar's true heritage is a fraternity begun by a mysterious (mythical?) man known as Christian Rosenkreutz (known as CRC).

The first trace of this fraternity was three centuries after the dissolution of the Templars. In 1614, a tract appeared called "Fama Fraternitatis, the Declaration of the Worthy Order of the Rosy Cross."[8]

This tract tells the story of how Rosenkreutz, a German nobleman, founded the Order in the fourteenth century, after journeying and studying occultism in the Middle East.[9] With his first four followers, CRC founded the "Fraternity of the Rose Cross." They built a headquarters called "The House of the Holy Spirit", where all the members gathered annually.[10]

THE VAULT OF CHRISTIAN ROSENKREUTZ

CRC died at the ripe old age of 150 because he wished to. Before his death, he fashioned his occult wisdom into a secret organization which would exist down through the centuries to save mankind. This society was secret because it had power to heal. He was entombed in a vault in the House of the Holy Spirit.

The tract's account claims that one of CRC's disciples discovered his tomb in 1604 and found strange inscriptions and a manuscript written in gold letters. Upon the door of the vault was an inscription which was interpreted as "In 120 years, I shall come forth."[11] Within the vault they found entombed a perfectly preserved body robed in the Rosicrucian vestments.

The discovery of CRC's tomb supposedly heralded a new era.[12] The members of this secret fraternity were called Rosicrucians. They claimed they never felt hunger, they had occult power,[13] and they also had access to lost secrets of science and medicine.

This is the nucleus of the Rosicrucian myth, much as the Hiramic legend is the kernel of Masonic legend. After this tract appeared, a "fad" of Rosicrucian literature began throughout Europe. Much of it was fraudulent. However, one item deserves our attention.

In 1616, a book called *The Chemical Marriage of Christian Rosenkreutz* appeared. Although allegedly narrated by CRC, its author was a scholar from Tübingen named Johann Valentin Andrae (b.1586).[14] The book is an occult allegory of a wedding in which some of the guests are killed and then brought back to life through alchemy. One of the key characters is a mysterious woman named Virgo Lucifera.[15] Her name means "Virgin of Lucifer."

The document is both an initiation and an allegory of alchemical transformation. Alchemy is central to understanding the depths of evil to which Freemasonry descends. Though alchemy is a complex subject; all we need to know for our purposes is that for both Rosicrucians and Masons, alchemy was the means to produce the philosopher's stone which gave them immortality. It would enable them to live forever. To alchemists, the eternal life they sought was actual physical immortality (a blasphemous parody of the eternal life Jesus offers).[16]

THE MATRIX OF MASONRY

Though the Rosicrucian fever died down, some still believed they were part of a fraternity of "Unknown Philosophers" known as the "Invisible College." Modern Rosicrucians have claimed Michael Maier, Sir Francis Bacon, Dr. John Dee, Wolfgang Amadeus Mozart, Benjamin Franklin, Thomas Jefferson and Sir Isaac Newton as alleged members.

Since the essence of being a Rosicrucian was total secrecy, the fact that some of these great men never gave any substantial evidence of membership in the fraternity proved little. Whether such claims are true or false, the idea of a hidden "college" of wise men with extraordinary powers behind the scenes has remained a compelling fantasy. Like most such fantasies, it has a dark side, and an element of truth. There were prominent leaders and thinkers who were professed advocates of Rosicrucianism.

The resemblances between Rosicrucianism and Masonry are evident. Both have their legends of masters entombed with secrets, and both promise immortality to members of their fraternity who labor diligently at their "Craft." Obviously, the matrix or mold into which modern Masonry was poured and shaped was the Rosicrucian order!

As Albert Pike explained the process:

> Resorting to Masonry, the alchemists there invented Degrees, and partly unveiled their

doctrine to the Initiates...by oral instruction afterward; for their rituals, to one who has not the key, are but incomprehensible and absurd jargon.[17]

When the occult "wisdom" of the Rosicrucians was joined to the stone Mason guilds of the 16th and 17th centuries, that was the beginning of the modern cult of Freemasonry.

One of the very earliest references to Masonry in English links the Lodge directly to witchcraft:

For we be brethren of the Rosie Crosse;
We have the Mason word, and second sight,
Things for to come we can foretell aright...[18]

It appears in a poem from 1638 by Henry Adamson of Perth called *The Muses Threnodie*. As most people know, witches claim to possess the "second sight."

As we shall see in the next chapter, most of the early Masons were indeed Rosicrucians. And that is only the beginning!

17

The Mother Lodge
and the Illuminati

Such well known Rosicrucians as Robert Fludd and Elias Ashmole were among the earliest and most prominent "Speculative" Freemasons.[1] Ashmole was maintaining contacts with the "Invisible College" which met at Oxford and included intellectual leaders such as Christopher Wren (the architect of St. Paul's Cathedral).[2]

Ashmole possessed five manuscripts of Dr. John Dee, the celebrated sorcerer who "brought through" the Enochian system of magic which now forms the bulwark of satanic ritual and demonic evocation.[3] Ashmole edited one of those manuscripts and became well known as an occultist.[4] He was initiated into the Masons in 1646.[5]

GUERILLA ARTISANS

The original "Free Masons" were stone masons who formed a guild to serve as a form of trade union. The increase in building cathedrals in Europe by the church in the Middle Ages created a real market for skilled stoneworkers.

The guilds provided assurance that its craftsmen were qualified. In those days, virtually everyone was illiterate, so "union cards" would have been worthless. However, since many masons travelled from city to city to work on cathedrals, it is speculated that the tokens of Masonry were used as ways of determining the level of the proficiency of a craftsman.

Nothing very ominous in all this. However, after the fall of Templary, some odd ideas began to be reflected in the work of the craftsmen—distinctly occult in nature. Look at any of the great cathedrals from the Middle Ages like Chartres or Notre Dame, and you will find masterpieces of their art. They are filled with occult symbols—demonic gargoyles, unicorns and other things which defy description. Why is this?

Although medieval Europe was Catholic, a large percentage of the common people were pagans, in practice if not in theory. Many of the feasts, saints, and even cathedral sites of the medieval church were actually pagan feasts, gods, and worship sites.[6] Notre Dame cathedral in Paris was built on the site of an important temple to the Horned God of Witchcraft, Cernunnos.[7]

Probably the majority of stone masons were pagans! They may have resented the intrusion of cathedrals onto their sacred sites. These masons could well have been "guerilla artisans." They were holders of the old religion and felt it was one fine joke to be paid by the bishops to build cathedrals which they (the masons) would then encrust with Witchcraft symbols! This is not speculation, for the results can be seen etched in stone!

PAGAN GROVES

Aside from the use of the "Green Man" motif, gargoyles, and other mythic beasts in the cathedral, there is a more central motif to these structures. One of the distinctive elements of "Gothic" architecture is the use of arched ceilings supported by pillars which stretch upward and interlock at the apex. Anyone who has seen these cathedrals cannot help but notice that these Gothic ceilings very much resemble trees arching over the worshiper.

The Bible speaks of "groves" where pagan worship was held (Exodus 34:13, Judges 3:7, etc.). Druids conducted their rites among groves of trees. Such groves were among the oldest pagan temples.[8]

The Israelites were strongly commanded to cut down the pagan groves of trees which were used for worship (Exodus 34:13). Thus it was a diabolical jest that these masons created an evocation in stone of this most ancient of idolatrous temples.

RIDDLES IN STONE

Notre Dame is noteworthy for another reason. High Masons realize that much of the symbolism in its ornamentation is actually a code which Masons used to pass on alchemical mysteries to their successors. These secrets are actually the very essence of modern Masonry![9]

The secrets could never be written, partly because they were "sacred" and were communicated only from mouth to ear, and because hardly anyone in the guilds could read. Much of what appears, especially on the central porch of Notre Dame, are actually memory aids which the elder Masters used to illustrate their secrets to the younger craftsmen.

They were riddles, etched in stone and intended to last for centuries! Most of these riddles cannot be deciphered without the keys which were provided in Masonic initiation. To the adept, the facade reveals the ultimate secrets of Masonry, the so-called "Royal Secret" contained in alchemical formulae.

An example is the Rose window in Notre Dame (Rosicrucian symbolism!). Even to this day, we have the expression "Sub rosa," meaning "under the rose." Something communicated "under the rose" was totally secret and could never be revealed.

The irony of this is that today, in America, if you were to ask a 32° Mason whose title is "The Sublime Prince of the Royal Secret" what the Royal Secret is, you would be greeted by a blank stare nine times out of ten.

THE STREAMS COME TOGETHER

Until the 17th century, there were only "operative Masons." However, in c.1600, Masonry evidently began initiating non-stonemasons into its ranks.[10] This formed the final catalyst which brought modern Masonry into being. Most Masons trace their founding to the first "Mother" Lodge which met in a tavern in London in 1717.[11]

In keeping with their Templar ties, these first Masons gathered on June 24, 1717, the feast of Saint John, a day held most sacred by the Knights

Templar,[12] which is also a high satanic holiday! In 1726, this lodge became "Grand Lodge of all England." This was followed by schisms between other "Grand Lodges," both in England and on the continent. In 1773, the second most influential Grand Lodge, the Grand Orient was formed in France.[13]

THE SEERS

This brings us to a most significant date in modern Freemasonry—May 1, 1776! On that date the final element in the evil equation of Freemasonry was introduced. We have already seen how strains of fertility cults, Islamic mysticism, alchemy, Templary and Rosicrucianism combined with the mason guilds of Europe.

Think of all these things as stones in an arch—a doorway into Witchcraft. These archstones needed a keystone to secure them. Satan had just the man for the job, and he would change the face of Masonry forever. The changes would be subtle, almost unseen, but they would harness together these exotic and ancient philosophies into a spiritual engine of enormous destructive power.

The keystone on the arch was provided by an obscure Jesuit-trained professor of canon law at the University of Ingolstadt in Bavaria, Adam Weishaupt. May 1, another high witch holiday,[14] was the date selected for the foundation of his secret society called the Ancient and Illuminated Seers of Bavaria (AISB for short).

It was founded on a mixture of Masonic secrets, Islamic mysticism and Jesuit mental discipline. The element which made it even more unique and dangerous was its scientific use of the drug, Alamout Hashish, to produce an "illuminated" state of mind. This was the drug of the Assassins.

Illumination had long been a cherished component of Masonry and other occult groups. The Masonic candidate requests, and is promised "light in Masonry." As he goes up the ladder of initiation, he receives "more light." It is because of this society's emphasis on illumination that the AISB became known by its more common title, the Illuminati.

ILLUMINATED MASTERS?

The lluminati are dear to the hearts of conspiracy theorists, frequently identified with the idea of a huge, shadow government intent on world domination. People are surprised to learn there are real Illuminati. The term is the plural of the Latin, Illuminatus, meaning "one who is illuminated." Thus, it means a person who has received the full extent of the initiation that is available through Freemasonry.

Technically, an Illuminatus is a Master Mason who has received all the "light" Masonry can bestow. He is beyond $32°$ and even beyond $33°$! Such people are known as Masters or Masters of the Temple, and collectively they are known by several names other than the Illuminati. Sometimes they are called the Great White Brotherhood or the Argentinium Astrum (Silver Star).

Whatever these people are called, they form an elite cadre of "super-Masons" with understanding of the principles of the Craft far beyond even the typical $33°$ Mason!

BIRTH OF A CONSPIRACY

One historian traces the Bavarian seers back to a 16th century Muslim cult of illuminated men called the Roshaniya in Afghanistan.[15] Again we see the pervasive influence of Islam in these secret orders. It may be from this Afghan connection that Weishaupt acquired his knowledge of hashish.

Weishaupt affiliated himself with the Masons, joining the Munich Lodge in 1777. He tirelessly worked to graft Illuminism into Freemasonry.[16] Weishaupt made it sound as though his society was working for noble ends like the freedom of mankind.

Through his use of drugs and occultism, Weishaupt produced an 18th century version of the Hashishim. His "illumination" was much more interesting than that offered by the regular Lodge. He played on the egotism which runs through Masonry and created a secret order within a secret order!

Many feel that Weishaupt's aim was to create a reign of genius "Philosopher Kings," with himself as the number one king. The highest degree of his order was that of the "Man-King."[17] He clearly believed in fostering controlled chaos, necessary for revolution. In many ways, the French Revolution and the Reign of Terror were very typical of Weishaupt's plans.

The anti-Christian hysteria of the French Revolution stood in marked contrast to its American counterpart. The Revolution's enthronement of a half-naked prostitute as the "Goddess Reason" on the altars of Notre Dame is a classic piece of Illuminist theater! The Revolution and its "Terror"

185

exemplify Weishaupt's view of humanity and the flow of history. In order to better understand the impact which Illuminism had on the sinister stew of Freemasonry, we need to look at his underlying philosophy.

THE LAW OF FIVES

The branch of the Illuminati into which I was inducted was supposedly directly descended from European chapters of the AISB. From them I learned something of the evil anthropology of the "Seers" of Bavaria. Did I not also possess the "sight?" Was I not part of a higher form of humanity? This is what I had been deluded into believing by Illuminism. I was told I was the next step up in the ladder of evolution, and that Illumined Ones were as far above human beings as people were above apes!

From somewhere, perhaps even the drug-ridden reaches of his mind, Weishaupt produced the "Law of Fives." His original inner council was structured around the pentagram (symbol of the Blazing Star, Sirius). According to our teaching, this inner council was made up of five men: Weishaupt's friend, Kölmer, Francis Dashwood (of the satanic Hellfire Club), Alphonse Donatién DeSade (where we get our word "sadism"), Meyer Amschel Rothschild (founder of the great banking house) and Weishaupt.

The number five is associated in magic with Mars. However, in Illuminism it is has even deeper levels of meaning. In an occult world view of any sort, nothing is regarded as coincidence. Everything has meaning. Therefore, it is highly significant that people have five fingers and toes, that the body has five appendages and there are five senses.

186

The most powerful images in Illuminist sorcery is the sign of Dagon (1 Samuel 5): the hand, palm forward with the five digits extended. This Law of Fives ruled history. Weishaupt taught that everything occurred in fives. Human history came in a cycle of five stages. Someone who understood these stages could manipulate history to his own ends. The five stages were:

I. **Chaos** (Verwirrung), the starting point of all societies, and the place of humanity in its "natural" state. It related, in Weishaupt's mind, to the goddess cults of antiquity, especially to the worship of goddesses such as Lilith, Eris, Diana or Kali.[18]

II. **Discord** (Zweitracht). Here, Weishaupt taught, a ruling class emerges and seizes control. This causes problems because the "average people" who are not on top resent the imposition of authority upon them and try to fight it. Weishaupt related this period to introduction (or imposition) of the worship of a male god (i.e., the God of the Bible or Marduk or Osiris).

III. **Confusion** (Unordnung). Weishaupt saw this period as a time when people would attempt to restore a balance between the two preceding forces. It is supposedly an attempt to reprogram human nature and make it fit into stage II. He related this period to the child-god, (Loki, Horus, etc.) or to a kind of devil.

IV. **Bureaucracy** (Beamtenherrschaft). The result of the synthesis of stage III failing. In this period, everything must be obsessively kept track of because folks can't take care of themselves. Weishaupt believed there would be a spiritual void during this stage, and absolutely no deity would be acknowledged. The only god becomes the ruling bureaucracy. Folks cannot abide this void and escape into fantasy, drugs, or madness. The rulers must continue to appear to control and know everything, and the burden on the slave class beneath makes them unfit for anything. They lose their jobs, end up on the dole or in hospitals.

It is during this phase that the destruction of the middle class takes place. Without a middle class to generate capital, the entire mess collapses into...

V. **Aftermath** (Grummet). This, Weishaupt taught, was the implosion of society into itself—a reversion to chaos. The bureaucracy crashes under the weight of its own red tape and things whirl out of control. Magic and nature now rule again, and the cycle is in preparation to begin again. Hence, the Scottish Rite 32° Motto: "Order from Chaos."

This is a lengthy dissertation, but in order to understand what Weishaupt's organization intended and what is happening today in the highest reaches of Masonry, it is essential for this five-stage theory to be summarized. It may not have a stitch of truth in it, and it certainly lacks the notion of a sovereign God, but we must realize that Weishaupt believed it to be true.

A BOLT FROM THE BLUE

The Illuminati's council truly believed they were riding a cycle of cosmic inevitability. Like a surfer, they just had to find the right point in the wave and catch it. By dovetailing his AISB with Masonic Lodges, Weishaupt had built the power base he needed. He felt the stage was set for the destruction of all the social institutions of the continent. Revolutionary France was very much an experiment in Illuminism.

Had God not intervened, all Europe might well have gone the way of France and the then-coming "Terror." Weishaupt's infiltration of Masonry might have been complete had not an AISB courier been blown off his horse and killed by a lightning stroke in 1785.[19] The courier was carrying papers written in the cipher of Illuminism, and dealt with the plans of the AISB to subvert the Masons and the governments of Europe.

The order was broken up by police and went underground. But no one really knows how far the AISB managed to infiltrate the Lodges, so today there is an immense amount of cross-pollination between the Illuminism and Freemasonry. Both the Grand Orient and the occult rites of Memphiz-Mitzraim (Egyptian Freemasonry) show influences of Weishaupt's hand.

It seems to have been Weishaupt who gave wings to the geo-political ambitions of the Masons in a fashion not seen since the Templars. Though Masonry always had its political overtones, Weishaupt's use of the Law of Fives, drugs and occult intrigues added momentum to the evil currents of the Lodge.

This final fusion of statecraft with sorcery created the Freemasonry we know today.

18

Albert Pike and the Congress of Demons

Soon after the percolating of Illuminism into the Lodge at the end of the 18th century that another extraordinary figure in Masonry was born. His name was Albert Pike (1809-1891), and his impact upon Freemasonry was as powerful as Weishaupt's.

While Weishaupt had to work from the outside, Pike was able to build upon his foundation and work within the "system." He became a Mason in 1850 and then, in a meteoric rise to power, was elected Grand Commander of the Southern Jurisdiction of the United States in 1859.

Arthur E. Waite quotes Dr. Joseph Fort Newton as saying that Pike "found Masonry in a log cabin and left it in a temple." He was the "master-genius of Masonry."[1] As mentioned earlier, Manly P. Hall calls him the "Plato of Freemasonry."

High praise indeed, especially for a man who

expressed contempt for Christianity, and who regarded Jesus as a teacher whose body is now dust. Today, some Masonic defenders are backing away from Pike since his writings, especially *Morals and Dogma*, are exceptionally nettlesome to anyone trying to prove that Masonry is a benevolent society which does not conflict with Christianity.

His impact cannot be denied, as he essentially made Scottish Rite Masonry into the institution it is today. The fact that, in addition to his Masonic titles, he also was the "Sovereign Pontiff of Lucifer" makes him someone to study very closely.[2] The evidence is that Pike regarded Lucifer as the true god. He, like Weishaupt, seems to have been a gnostic and a Manichæan—a dualist at the very least.

Since Pike chose to follow the mystery religions of Ba'al, he turned his back on God. If we look at his writings and statements attributed to him, we find that he acknowledged Lucifer as the true god and Adonay (the Biblical God) as the god of evil.[3] Pike's "fruits" are definitely satanic in nature. (Matthew 7:15-20). At some point, he moved from being a typical Mason. He received "more light" and decided to cast his lot with Satan.

BAD COMPANY!

Pike seems to have been definitely influenced by the Italian Freemason and revolutionary, Giuseppi Mazzini (1805-1872). Mazzini and Pike were the malignant "Bobbsey Twins" of 19th century Masonry, with Pike running the show in the States and Mazzini in Europe. Both were military men with noses for rebellion. Pike was a general on the side of the Confederacy in the Civil War, even

191

though he was a "Yankee" born in Boston![1]

Mazzini had formed a society in Sicily called the Oblonica, which translates loosely as: "I reckon with a dagger." As is typical with the Masons, Mazzini formed an order within an order. This elite inner group was called by a term much more familiar to the reader—the Mafia!

Although most people know what the Mafia is, few people realize it was founded as a Sicilian Masonic terrorist organization. The name Mafia emerged around 1860 and is an acronym for **Mazzini autorizza furti, incendi, avvelenamenti**—Mazzini authorizes thefts, arson and poisoning.[4]

Illuminati touches are also evident in the Mafia. Remember the Illuminati Law of fives and their sign—the hand with palm forward and five digits extended? Some people may remember that the Mafia was known by another term, Il Mano Nigro—the Black Hand! In the society's heyday, Mafia crimes were often sealed by a black handprint at the scene, as if someone had taken their palm, pressed it in ink and then made a hand print on the wall.

The Mafia also has its blood oaths, its code of silence, the Omerta, and it does "take care of its own." It is an ideal Masonic organization.

Pike, on the other hand, helped create what I call "Masonry in percale," the Ku Klux Klan! Pike, the old Confederate general, was a wily strategist who knew that if he could leave behind a secret terrorist society in the south to fight against freedom for black people as a rear guard action, the south's defeat might not be in vain.

Although these facts may stun Masons, the

Lodge has always been racist! Almost no black men were ever admitted to Lodges because of the qualifications that the candidate be a "man, *free born,* of good repute and well recommended."[5] This is what the term "free" in Freemasonry stands for. You must be a "free and accepted Mason."

This rule kept blacks of slave ancestry out of the Lodge up to the last decade or so. Black men were forced very early in U.S. history to adapt their own form of Masonry which is called Prince Hall Masonry.

AN UNHOLY ALLIANCE

In 1870, Pike and Mazzini completed an agreement to create a supreme, universal rite of Masonry which would over-arch all the other rites, even the different national rites.[6] It would centralize all high Masonic bodies in the world under one head. This head would be, in the ultimate sense, Lucifer. However, Pike and Mazzini would be his human regents. To this end, the Palladium rite was created as the pinnacle of the pyramid of power.[7]

Domenico Margiotta, a 33° Mason, has written:

> Palladism is necessarily a Luciferian rite. Its religion is Manichæan neo-gnosticism, teaching that the divinity is dual and that Lucifer is the equal of Adonay, with Lucifer the God of Light and Goodness struggling for humanity against Adonay the God of Darkness and Evil...Albert Pike had only specified and unveiled the dogmas of the high grades of all other masonries, for in no matter what rite, the Great Architect of the Universe is not the God worshipped by the Christians.[7]

Please note that last statement carefully! The Luciferian doctrine, we are told, is implicit in the lower degrees, and only becomes an explicit teaching in the highest degrees. The highest of the high was the Palladium.

It would be an international alliance of key Masons. It would bring in the Grand Lodges, the Grand Orient, the 97 degrees of Memphiz-Mitzraim (the Ancient and Primitive Rite) and the Scottish Rite. The name, Palladium, was taken from a Masonic order founded in 1720 which died out, only to re-emerge in Charleston under Pike.

Although some Masonic apologists like Arthur Edward Waite claim there was no Luciferian Palladium under Pike, their protests ring hollow. Waite himself was a sorcerer and taught people through his books how to conjure up demons and sell their souls to the devil!

Though he kept pretending to be a "white-light" Christian magician (??), he wrote books with titles like *The Book of Black Magic and Pacts*. How gullible do they think we can be? One can no more be a Christian magician than a Christian crack peddler!

I was brought into Palladium Lodge (Resurrection, #13) in Chicago in the late 1970's and received the degree of "Paladin" in that Lodge in 1981 from the son of one of the leading occultists in the late 19th century—an associate of Aleister Crowley. Evidently there was (and is) Palladium Masonry being worked in the 20th century.

I am ashamed to admit it, but I, myself, stood in Lodge and joined in the traditional Palladium imprecation, which is (translated from the French): "Glory and Love for Lucifer! Hatred! Hatred! Hatred! to God accursed! accursed! accursed!" Although I cannot be absolutely certain, I have no reason to doubt the authenticity of the initiation I received. It certainly came right from the pit of hell!

DRIVING THE ENGINERY OF EVIL

What made the Palladium rites so distinctive? What made them the glue that held all these different forms of Freemasonry together? One of the arguments raised against the idea of conspiracy in history or religion is the simple fact that most people cannot sit down and agree on the color of parking meters in a town, much less the destiny of the world.

This point is well taken, and that is the purpose for the Palladium being revived. It took the fuel of Weishaupt's dangerous vision for humanity and connected that energy to the pistons of Freemasonry. Those pistons were then driven in rhythm by the Palladium.

There are basically two ways to get people to move in agreement. One way is through force, either military or of a moral or legal nature. The other way is through a spiritual transformation. If you can somehow plug everyone into the same spiritual power source, if they surrender to that power, and if that power is not concerned with free will, then a remarkable kind of unanimity can be achieved. This can be seen to a small degree in what is called "mob psychology."

In a greater degree it can be seen with Hitler and Khomeini. Many Americans were astounded and dismayed at the fervor and persistence of the anti-American demonstrations during the Iran hostage crisis. Guns were not being held to the heads of those hundreds of Muslim fundamentalists who daily hurled their tirades at the "Great Satan." Similarly, the old newsreels we watch of Nazi rallies are frightening in the lockstep of the participants.

It is evident that both Hitler and Khomeini were plugged into some sort of malignant energy and somehow harnessed that energy into the people who followed them. They exercised amazing control over thousands of people who had submitted to them as spiritual/national leaders.

Now turn the power knob up still higher, and you will understand how the Palladium worked.

"THE MIGHTY DEAD"

What could you offer to one who already had reached the pinnacle of Masonry? Many, if not most, already had wealth and power. The only further enticement of any power would be an ancient one:

> Ye shall not surely die...ye shall be as gods,
> knowing good and evil.   Genesis 3:4-5

Such men would be offered immortality and godlike wisdom. In common with most other gnostic-type cults, the Palladium taught that the serpent told the truth in the Garden of Eden. Basically, the candidates were exposed to a five-step program in the Palladium.

1. Adoption: This is where the Mason is brought into the "fellowship" of Lucifer. He is guided into swearing an oath and being yoked to the temple of Lucifer. Ultimately, he is led into making a pact with Lucifer. This is basically "selling of the soul" to the devil. The Mason promises to surrender himself, body, soul and spirit to Lucifer, usually for a period of seven years. In return, Lucifer promises to grant him all his worldly desires. After the seven years are up, if he has been a good servant, Lucifer will give him another seven years. If he has failed in some fashion, his life is taken.

2. Illumination: through drugs and occult techniques of the Seers, the so-called Third Eye would be opened, not just partially (as in psychics) but completely. This "eye," also called the Ajna chakra, is felt to be the point of contact between humans and Lucifer-consciousness. It is supposedly located in the forehead, above and between the two visible eyes.

To "open" the eye a little bit is to experience psychic powers. To open the eye completely is to have your brain flooded with the "pure" consciousness of Lucifer himself. This is why one of the Masonic symbols is the "All-Seeing Eye." It is a symbol of Illumination.

This is Satan's counterfeit for being Born Again. In it, you acquire a "personal relationship" with Lucifer. You begin to think his thoughts and see with his eyes. You begin to look at humans the way he does. It is not a pretty experience!

3. Conversation: This involves communicating with "The Mighty Dead." Both my personal experience and historical testimony concur that spiritism (communications with the dead) plays a significant part in the Palladium. Mediumship is encouraged, and "conversation" with the sages of history is vital.

We spoke (allegedly) with such luminaries as Jesus, Plato, Francis of Assisi, King Arthur, the emperor Nero, Aleister Crowley, and even Hitler! We were very ecumenical! These "wise and powerful beings" gave us advice and taught us how to bring our bodies and wills more perfectly under the subjection of Lucifer.

4. Congress: The initiate would be led into being literally "married" to these dead. Usually this was done either by having a medium of the appropriate gender be possessed by the "dead spirit" (actually a demon) and then a sacrilegious wedding ceremony was consummated. It was believed that the magical "virtue" of the spirit would flow from the possessed medium into the initiate through the act of intimacy.

Sometimes an actual spirit would be invoked through what is called an VIII° working (magical self-abuse) in the hope that a succubus ("female" demon) or incubus ("male" demon) would manifest. The idea was that the wisdom and god-like power of the spirit being would gradually, through repeated congress, totally infuse the mind, body and will of the initiate.

Once this awful goal had been achieved, the initiate is brought into:

5. Union: At this point, the soul of the initiate is totally eclipsed by the evil spirit. In other words, there is virtually "nobody at home" except the demon! This is known as "perfect possession" and takes many years of inviting the demons to come and own the person. At this point, the initiate ceases to be an autonomous individual. He is but a fleshly "glove" with a demonic "hand" inside controlling his every move.

Since the Lord has shown that Satan's kingdom is not divided against itself (Matthew 12:25-26), we can assume that his demons are willing to cooperate to serve his ends. This is especially true since Satan is not a kind master and can be unbelievably cruel in his punishments for disobedience.

Men and women who have made it to stage 5 (there are not large numbers of them) will be so utterly given over to the will of Satan that they become "satanic saints." They will go anywhere and do anything to please their master.

It is through people like this that Weishaupt's "Law of Fives" can be utilized. The leaders observe what they perceive to be turns in the political cycle and have their "saints" ready to move in key areas, to act without compassion or any shred of humanity. Such "illuminated beings" regard humans the way we regard cattle. A famine here, a pogrom there—what are a few hundred thousand human lives compared to the noble cause of the Great Architect?

What sort of people could precipitate the mass murders in Stalinist Russia or the Nazi holocaust? Who could force a politicized famine upon the people of Ethiopia? These are the work of men and women (or their minions) who are the super-rich, international, geo-political equivalent of Charles Manson or Jim Jones! These are people who aren't really people anymore. They are cosmic terrorists—puppets controlled by demons.

Because they are puppets, they move in unison, with an end in mind which is transgenerational. Though they will die someday soon, the demonic masterminds which control them will live on, and assure themselves of new "houses" in which to dwell (Matthew 12:43-45). These demons don't care if their goals take generations to achieve. They think time is on their side.

What upwardly mobile young banker, lawyer, politician or even preacher would not be willing to invite a "mighty being" who was wise and immortal into themselves for a consultation? Once the "wise one" who is the source of such "ancient knowledge" has been invited in, he is an increasingly difficult guest to evict. The person becomes literally addicted to the power and influence of this new "friend."

With the development of New Age and "channeling," Satan has managed to begin the mass marketing of what was reserved for the elite. Like spiritual "junkies," these people have submitted to increasing levels of depravity because they have been promised that union with these demonic beings will give them immortality and enable them to evolve into gods through alchemy and black magic.

The metaphor used on me was this:

BREAKING IN A NEW PAIR OF SHOES?

I was told that each time one of these "mighty beings" came into me, it would stretch my spirit, like breaking in a new pair of shoes. Each time a person wears the shoes, they conform to the shape of his foot. Similarly, each time the demon enters the initiate, his etheric (soulish) body "grows" to fit the god-being a bit more.

Different demons would come into me at different times—at first as "spirit guides," but later they claimed to be high level masters or even gods. I remember the first time a supposed god-being tried to possess me. The energy was so intense that my body broke into profuse sweats. My form actually expanded to the point that I physically tore the seams completely out of the robe I was wearing. This was in the presence of my wife, so it was not a hallucination! This stuff is frighteningly real!

After almost ten years of practice, I was able to accommodate the same sort of high level being with hardly anything more serious than an accelerated heart rate. I had "grown" into them (or at least was being led to believe that such was the case). The only permanent effect was the premature greying of my hair, which happened virtually overnight in my mid-twenties and then spread progressively.

By the grace of God, I was only permitted to make it to the fourth level of this terrible process before He intervened through the prayers of a Christian woman. Although the blood of Jesus can cleanse from all sin (1 John 1:7), I'm not certain how many people would be spiritually salvageable once they got far into step five.

However, I knew what it was like to have my mind filled with the scalding hot fog of the "brilliance" of Lucifer. I experienced "more light" in

Masonry with a vengeance! I knew the feeling of being linked mentally into a vast spider web of communications, and being part of an invisible army of slaves almost totally at the command of the Deceiver himself.

Even the typical Mason experiences (to varying degrees) the first two steps of this hellish progression: Adoption and Illumination. He is adopted into Lucifer's family when that cable-tow is placed about his neck, and when it is removed at the end of the first degree oath! That oath placed him in bondage to Lucifer, whether he knows it or not.

Similarly, the "illumination" experienced when the candidate's blindfold is removed at the end of the oath is intended to shock him (along with the sound of all the Masons in the temple clapping their hands in unison) into an "altered state of consciousness." Though it doesn't seem to work that well in most cases, it is intended to open spiritual doorways for later evil.

Thus we see that even a simple Mason has placed a foot upon the path towards the congress of demons. How far they progress in that path depends upon a large number of variables, not the least of which is the number of people praying for the man, and his Christian background (or lack thereof). I was not a Christian when I was initiated. However, over the years, the Lord raised up people to pray for me.

I was not the typical Mason. I joined the Lodge at the suggestion of my mentors in the occult because they felt the Lodge degrees would be an essential part of my spiritual "evolution." I did this obediently, and tried my best to be a faithful Mason.

I was instructed to take the York Rite path at first. Once I entered the Knights Templar (the equivalent of the 32° in Scottish Rite), I was considered "worthy" of the European Masonic degrees. My progression through those degrees was fraught with strange initiations too awful to recount. In a gradual deadening of my conscience, I was brought through the Egyptian rites of Masonry and received the 90°, a level few U.S. Masons are even aware of!

At the same time, I was working hard in the regular Masonic bodies. I was seeking to serve, and they provided opportunity. I held offices in virtually every Masonic body except the Scottish Rite and the Shrine. Masonry had become an important part of my life, especially since I had come to know the real secret of the Lodge through my occult involvement in the high level offices.

Along the way, I did meet a couple of other "pilgrims" who were deeply involved in esoteric Masonry, but most Masons were blissfully in the dark about the "light" they had. That is where their leaders want them. They didn't even know the secret they were guarding so carefully.

Like workers in a classified project, they were only allowed to know as much as they needed to know to work in the system. These poor men, many of whom were church-goers, were ignorant of the big picture, and would be appalled to learn the truth concealed behind the layers of allegory. We are going to lift the veil off this "sanctuary" and expose the darkness of Freemasonry's light!

19

The Witchcraft Connection

It is vital to understand that the past interchange between Masonry and these various occult groups did not stop in the 18th century. If anything, it has grown more prominent in the past century.

There is something about the Lodge that has always attracted sorcerers. The list of occultists and witches in the last century who were Freemasons reads like a Who's Who of 20th century occultism:

- **Arthur Edward Waite**—occult writer and Masonic historian

- **Dr. Wynn Westcott**—member of the Societas Rosicruciana in Anglia and founding member of the occult Order of the Golden Dawn.

- **S. L. MacGregor Mathers**—co-founder of the Golden Dawn,

- **Aleister Crowley**—master satanist of this century and founder of the anti-christ religion of Thelema.

- **Dr. Gerard Encaussé**—(Papus) masterful author, teacher of the Tarot and leader of the occult Martinistes society.

- **Dr. Theodore Reuss**—head of the O.T.O., a German occult/satanic society which made Crowley its head for the British Isles.

- **George Pickingill**—the master warlock (male witch) of 19th century England, leader of the "Pickingill covens."

- **Annie Besant**—leader of the occult Theosophical society and Co-Masonic hierarch. (Yes, there are female Masons!)

- **Alice Bailey**—founder of the New Age organization, Lucis (formerly Lucifer) Trust.

- **Bishop C. W. Leadbetter**—Theosophist, mentor to the failed New Age "Christ", Krishnamurti, and prelate in the occult Liberal Catholic Church.

- **Manly P. Hall**—Rosicrucian adept, author, founder of the Philosophical Research Society.

- **Gerald B. Gardner**—founder of the modern Wiccan (white Witchcraft) revival.

- **Alex Sanders**—self-styled "King of the Witches" in London and one of the most influential leaders of Wicca after Gardner.

Would you really wish to belong to an organization which welcomed these powerful sorcerers into its midst with open arms?

In addition, there are the many minor occultists (as I was), who are in the Lodge, drawn by its mysterious power. At least three or four of my male witch friends were in the Masons, and all my leaders were!

There is a real reason for this strong affinity between Masonry and witchcraft. It is because the Lodge is plugged into an international network of witchcraft, a hierarchy of evil.

Recognizing the All-Seeing Eye as an occult symbol, its use on the Great Seal of the United States is not without significance (see the back of any dollar bill). You will note that the Eye is perched atop an incomplete pyramid with the date 1776 A.D. in Roman numerals at the bottom.

Remember 1776 is also the year Weishaupt founded the Illuminati! The trapezoid (what the unfinished pyramid really is) is a most significant symbol in Satanism.[1] The symbol on that seal is actually a metaphor for the oppressive hierarchy which reigns over the Masonic Lodge, and over much of U.S. government. The pyramid is something like the illustration on the next page.

Being a Mason (of whatever degree) is like going through your life with all that spiritual garbage weighing down on you. It is like having a King Kong-sized monkey on your back! While all levels of Masonry have their share of witches, the Palladium, the Illuminati, the Ancient rites and the Supreme Council are especially likely to have them, in one form or another.

The Mason is "unequally yoked" together with all these unbelievers and witches in rebellion to the Word of God (II Corinthians 6:14-18). That alone is enough to knock the spiritual stuffing out of any man!

Lucifer:
"The Light of Limitless Nothingness"

T.G.A.O.T.U. .
(Ain Soph Aur)

The "Seven"

The 9 "Unknown Men"

The "Illuminati"

The Palladium

European
(Esoteric)
Masonry

Ordo Templi Orientis (O.T.O)

Ancient & Primitive Rite (97 Degrees)

The Order of the Trapezoid

Supreme Council of Grand Sovereign Inspectors General

Grand Sovereign Inspectors General – 33rd Degree

(The Shrine)

Scottish Rite or York Rite Masonry

Blue Lodge Masonry

U.S.
Masonry

The Masonic Hierarchy

A SPIRITUAL PYRAMID SCHEME!

Freemasonry is like the fabled "pyramid scheme." It is a hierarchy in which the highest levels essentially leech off the lower levels. Just as in the marketing schemes, the person at the top of the pyramid draws in most of the revenues because of the efforts of hundreds or thousands of people under him.

The same element works within the Lodge, and,

to a much smaller degree, in a witchcraft coven!

This works in three ways. First, it is a financial pyramid. As already mentioned, a Mason must spend hundreds of dollars, perhaps close to a thousand, to go through the degrees. Additionally, he must pay dues every year to each and every body he has joined. Depending on the level of involvement, this could amount to several hundred dollars a year.

While some of that money goes into necessities, and some into charity, some also ends up in places of which Lodge members have no knowledge. Our local leaders were obviously not getting rich, but there was a lot of free-floating cash somewhere up in the ranks.

Nowhere is this more evident than in the "rich-boy's" organization, the Shrine. The public impression of the Shrine is that of a philanthropic organization, and all the millions of dollars it raises through Shrine circuses are funnelled into its hospitals. However, it recently came to light that less than 2% of the money raised at all these high-visibility events actually went to Shrine hospitals. These hospitals are already fully endowed, and have large surpluses.

WHERE DID ALL THE MONEY GO?

Most went directly into the huge temples and country clubs the Shrine runs. Aside from this, remember all the "fun" things the Shriners do, the little cars, the Oriental band, the camel brigades, and on and on! All those things cost money, some of them, lots of money! Much of this money is coming out of the pockets of "good Christian men," while missions organizations go begging all over the world!

THE ISLAM CONNECTION

Throughout this book, there have been allusions to the frequent historical links between the Muslim religion and the origins of Masonry. The most obvious connection with Islam in the world of the Lodge is, of course, the Shrine. Most don't know that the actual title of the Shriner organization is the "Ancient Arabic Order, Nobles of the Mystic Shrine."

Not only is the Shrine openly evocative of Arab culture, the "shrine" is actually the sacred shrine of Islam, the Kaaba in Mecca! Few outsiders realize that behind its exotic, clownish exterior, the Shrine ceremonial is steeped in demonic Islamic religion, utterly foreign to the God of the Bible!

For example, the Shrine initiate must swear the usual awful oath on the holy book of the Islamic faith, the Koran, in addition to the Bible, ending thus:

> ...and may Allah, the god of Arab, Moslem and Mohammedan, the God of our Fathers, support me to the entire fulfillment of the same, Amen, Amen, Amen. **(Shrine Ritual Monitor, Allen Publishing, pp. 35-39)**

The Shriner is swearing in the name of Allah. Contrary to popular belief, the Islamic Allah is not just another name for the true God. He has no more resemblance to the God of the Bible than does the "Great Architect."

History shows that before the "prophet" Muhammed elevated Allah to special status, he was essentially a second-rate little rock of an idol inside the pagan shrine, the Kaaba. He was one of a crowd of some 365 little "rocks" in there, and was the tribal god of Muhammed's tribe, the Quraish.

Yet today, this "rock" is worshiped by 600 million Muslims! Like any other idol, there is a demon behind it! (Leviticus 17:7, Deuteronomy 32:16-17). That demon draws energy from being worshiped for his master, Satan! It is certain that this "Allah" draws a great deal of foul satisfaction from watching good "Christian" men kneel down and swear allegiance to him!

The Muslim denies the deity of Jesus Christ and His resurrection, to say nothing of the gospel of grace. Islam denies all the cardinal fundamentals of the Christian faith. You cannot be a faithful Christian and call Allah the "God of our fathers!" The Shriner is calling a demon named "Allah" his god!

CELESTIAL PROSTITUTES

People often marvel at the high level of immorality which surrounds Shrine conventions. One of my female friends was highly offended when I told her I was joining the Shrine. She told me that whenever the Shriners were in town, she was continually harassed, and even physically accosted in an improper manner by them.

Shriners are exhorted to regard their red Fez as an analog to the white apron and to never do anything while wearing it that would bring shame to their mothers. They either have very strange mothers, or their consciences have been completely numbed!

By submitting to the god of Islam, they have

come under the power of a religion that has a strange view of the afterlife. Both in this life and in the next, women are viewed as property. Most people know that Muslims are polygamous, and that Islamic women in strict Muslim nations have virtually no rights. They must be covered from head to toe, and are basically treated as if they have no souls or identity of their own.

In the Islamic paradise, the men are waited on by houris-beautiful, angelic women. These women are mindless and exist only to serve the sexual needs of the Islamic "saints." They are basically celestial prostitutes! If this Islamic spirit has come upon the Shrine "Nobles," no wonder they behave like over-sexed adolescents in these conventions.

In the "Jesters" club, an elite group within the Shrine, women have reported to us that their husbands abused them physically and sexually and even share them with other Jesters, like a wife-swapping club. If this is true in the larger Order, it could explain the horrible Islamic attitude toward women which basically regards them as objects.

With or without the immorality, the Islamic powers which loom over the Shrine should make it a place any Christian would flee.

The red Fez itself, associated with the Shrine, is actually an article of ceremonial attire among Moroccan Muslims. Its color is said to reflect the fact that centuries ago, Islamic armies invaded Fez and slaughtered thousands of Christians who resided there. The blood of Christian martyrs ran in the streets, and the Islamic "holy" warriors dipped their headgear in the blood and dyed them scarlet. Thus, the Fez is a commemoration of the murder of thousands of Christians! No wonder Satan smiles when Christians wear it proudly!

How is Jesus glorified by men who are supposedly His disciples riding around on funny little expensive motor cars in parades wearing evil red hats, while millions of people have never heard about the good news of salvation through the cross?

I'm not trying to be a spoil-sport here, and I like a good time as much as the next guy. However, this should be done only after your spiritual obligations have been met! It is estimated that the average Christian in the U.S. gives 3 cents a day to missions. That is certainly less than the typical Shriner spends on his activities! Every missiologist I have read would agree that the U.S. is far behind in its support of foreign missions. We are nowhere near being tithe-payers!

I'm not trying to be a legalist. We are not saved by tithing. However, if the vast majority of Christian men aren't paying a full tithe, (and they are not) there is something wrong. This is literally draining millions of dollars a year out of the church at a time when many Godly missionary organizations (both church and parachurch) are barely able to keep afloat! What a fine trick on the church by Satan!

PSYCHIC VAMPIRISM?

In the occult, we used to talk about psychic vampires—people who seemed to suck the life out of a person. Of course, black magicians excel in this. They leave people feeling drained. What most people don't realize is that an organization can function in much the same way.

212

The Lodge functions like a spiritual sponge in many ways. Think of all the millions of man hours Masons put into their Lodge work: memorizing the degree material necessary for advancement, attendance at meetings, extracurricular lodge activities (dinners, banquets, funerals, picnics). Those Masons who are Christians pour hours of time and energy a week into the Lodge, and it just laps it up and begs for more.

I know, I used to be heavily involved in Lodge work. I was out of the house at least two week nights! Then, being a Lodge officer, I spent additional hours memorizing the ritual work. I had to be there before the Lodge opened and after it closed. I had to attend all Lodge functions, especially funerals.

Think of the Lodge meeting itself. It is opened in solemn fashion, with a ritual which may take fifteen to twenty minutes. If there is an initiation, the meeting can run two hours, sometimes three or four hours for third degree. All that energy is going somewhere, friends, and it isn't to God!

I can only speak from witchcraft experience, but quite often our leaders would suck the energy right out of us. They were accomplished psychic vampires, whether they realized it or not. Someone, somewhere, is getting an awful lot of energy out of these thousands of Lodge meetings. Ultimately, of course, it is Lucifer, who is delighted to receive it as worship!

This is energy not being expended in Godly church activities. These men could be teaching Bible studies, running youth groups, visiting the sick or doing neighborhood witnessing. But no, they are sitting in a Lodge room watching ancient and dusty mummery being performed while the light of the Holy Spirit within them flickers out.

Over and over, we see Christians who don't see the trap, join the Lodge, and then gradually it sucks all the life out of their walk with Jesus. It banks their fires of zeal and turns them into dead backsliders. Some stop going to church.

This may not happen to all Christian Masons, and if it doesn't, it is only because of God's mercy. The Holy Spirit will not continue to bless a man who continually sups at the devil's table (1 Corinthians 10:21, Genesis 6:3). Sooner or later, something will give. Sadly, it is often the church activity.

THE IMAGE OF JEALOUSY!

The Masonic temple is a temple of witchcraft! There is little doubt about that. Veiled within its symbols are the deities and even the working tools of witchcraft!

As we have seen, the square and compasses are representations of the generative organs—the "sacred altar" of witchcraft! The blazing star at the center of the Lodge is the witch's pentagram, symbol of the god of Satanism, Set! The Letter "G" stands for generativity, sexual potency.

The apron is a tool of the ceremonial magician and a symbol of Lucifer's priesthood. It also functions as a sacrilegious parody of the veil in the temple of Solomon, in that it veils the "most holy place"—in this case, the Mason's genital region.

The Masonic initiate and the witch initiate both

214

stand at the northeast corner of their respective temples. Both are stripped of all metal (for occult reasons), both are blindfolded and tied with a cord. Both are challenged at the door of their temples by sharp instruments piercing their naked breasts. Both must swear dreadful oaths to keep secrets they do not know yet. Both are "illuminated" at a key point in their initiations after swearing the oaths by having their blindfolds removed.

The resemblances are manifold and striking and should chill the bones of any Mason. Some have speculated that these similarities exist because the founders of 20th century white witchcraft (Wicca) were all Masons and, therefore, the Wicca rites are copies of Masonry.

If that is true, it leaves the Mason in just as dismal a spiritual place. It means he is spiritually connected to an organization whose rites can effortlessly slide into witchcraft and devil worship and fit beautifully! If Freemasonry is so Godly, how could it possibly be interchanged by both witches and satanists so freely?

Virtually all the above mentioned resemblances are part of the ancient practices of pagan antiquity as well. Witches 2,000 years ago were doing the same things that Masons are doing today. Masonic writers boast of this (although they don't use the word "witch," they talk about "mystery religions," but it is the same thing).

Let's face it, the Masonic tie tacks and rings which so many Masons wear proudly to their churches on Sunday are sexual idols. The true God of the Bible is not a sex organ! That may seem a ridiculously obvious statement, but the Mason needs to be reminded of it. This is the very "image of jealousy, which provoketh to jealousy" (Ezekiel 8:3).

The gods of the pagan nations around Israel, like Ba'al, were all sexual idols! This is precisely what God does not want in His church, and yet, Masons are flaunting both their idols and their memberships.

It is a testimony to the graciousness and lovingkindness of Father God that these churches are not flattened by the breath of His nostrils, that they are not "vomited out" of His mouth (Revelation 3:16).

However, both they and their individual members may well be paying a horrible price for their continued winking at the sin of Freemasonry in their camp!

20

Masonry's Trap Closes!

After I came out of Masonry, I was astounded by one of the first scriptures the Lord showed me with its uncanny resemblance to what is happening today in Lodges all over the world.

After seeing the "image of jealousy," Ezekiel is taken in a vision into the temple in Jerusalem. Within the temple walls, the prophet was shocked to see idols painted on its inner corridors, with seventy of the "ancients" of the house of Israel worshiping them with incense. The Lord tells Ezekiel:

> Son of man, hast thou seen what the ancients of the house of Israel do in the dark, every man in the chambers of his imagery? for they say, The LORD seeth us not; the LORD hath forsaken the earth.   Ezekiel 8:12

Then the prophet is taken further, and sees the women weeping for the "dying and resurrected" god, Tammuz (Hiram Abif?). The final horror in the eyes of God, however, was the following scene:

> And he brought me into the inner court of
> the LORD's house, and, behold, at the door
> of the temple of the LORD, between the
> porch and the altar, were about five and
> twenty men, with their backs toward the
> temple of LORD, and their faces toward the
> east; and they worshiped the sun toward the
> east.   Ezekiel 8:16

That is precisely what happens at every Lodge
meeting! Although the Masonic temple is supposed
to represent Solomon's temple, it is facing exactly
backwards from the way the real temple was
situated!

The Lord deliberately had the temple rites facing
west, away from the rising sun, which was the chief
symbol of the pagan gods of the time. To face the ark
of the covenant, one had to face west. In the Lodge,
all devotions are made facing east!

Not only that, the Masonic ritual boldly
proclaims that:

> As the sun rises in the east, to open and
> govern the day, so rises the Worshipful
> Master in the east, to open and govern his
> Lodge, set the craft to work and give them
> proper instructions.[1]

Here is a direct identification of the "Master" of
the Lodge with the sun god (Ba'al). How can
Masonry be an enactment of the rites of Solomon's
temple when the whole thing is "oriented" the
wrong way! This is a similar concept to the satanic
Black Mass, where all the Christian symbols like the
cross or the Lord's Prayer are inverted or said
backwards.

THE PLACE OF DARKNESS

When the new Mason sees the Lodge, he will note there are officer's stations in three of the four compass points: the Master in the east, the Junior Warden in the south and the Senior Warden in the west. The north alone has no dignitary. In the lecture, the candidate is told:

> Because this and every other Lodge is, or ought to be, a true representation of King Solomon's temple, which was situated north of the ecliptic; the sun and moon, therefore, darting their rays from the north. We therefore, masonically, term the north a place of darkness.[2]

That should be an interesting remark to anyone familiar with the Bible! First of all, the temple's holy place and most holy places were not dependent on the light from the sun or moon. No, it was lit by the seven branched candlestick (Exodus 25:31-40, 26:35).

More importantly, the Bible teaches that the north is the dwelling place of God. God is enthroned in the north (Isaiah 14:13, Ezekiel 1:4, Psalm 48:2). How interesting that the Mason must turn his back on the true temple of God to worship, and that he also regards the place where God's throne sits as the "place of darkness."

You see, the trip wires are everywhere! Spiritual traps await the Mason around every corner. Hardly a facet of the ritual or doctrine of the Lodge exists which does not hold some blasphemy, some hidden peril! The bottom line for the Christian Mason are these questions:

- How many things need to be wrong with an institution before you will renounce it?

• How many sins does it have to lead you into before you say it's enough?"

For a sincere Christian, the answer should be, "Just one." We have shown you dozens! You need to understand the spiritual dynamics of these traps, and how cleverly all Masons have been duped by the master Deceiver himself—Satan.

WELCOME TO THE WORLD'S LARGEST COVEN!

Masons are adopted into the Masonic order. The word adoption is key here because it is a deliberate copy of the sacred act which takes place when a person becomes Born Again. You are adopted into God's family (Romans 8:15, John 1:12)!

In Masonry, you are bound over into slavery to pagan principalities and powers! When the new Mason has sworn his solemn oath, the Lodge Master orders the Senior Deacon to remove the cable-tow or cord from his neck, "as we now hold the brother by a stronger tie."[3] Like the new witch, the Masonic candidate is brought in bound with a cord.

In a limited way, the cable-tow is explained to be a symbol of the authority of the Master. The Mason is to respond to commands given by the Master if "within the length of his cable-tow."[4] However, the deeper meaning of this cord, which is given in witch initiations, is not explained to the Mason.

In the oral tradition of witchcraft to which I belonged, we were taught that the cord symbolized the umbilical cord which connected the goddess of witchcraft, the "Great Mother" or "Queen of Heaven" (as the Bible calls her) to her hidden children, the witches.

220

In an overly literal parody of the teaching of Jesus on being Born Again, the cable-tow represents the cord of life which ties the new Mason to the gods of witchcraft! It binds the Mason as a "Born Again pagan," a "hidden child" of the goddess and her horned consort, Lucifer!

For the Christian Mason, this creates incredible spiritual tension. He has been made a child of God, and now he has also been adopted into witchcraft! Talk about being a *double-minded man* (James 1:8)!

In witchcraft, the cord has a more mundane meaning. It is used for bondage, both physical and spiritual. If a witch wishes to prevent someone from doing something, they make a puppet or doll of the person and then bind it up with their ceremonial cord. Similarly, the Mason who is bound three times in the three degrees, becomes snarled in occult energy coils of the "old serpent" like a kitten in the grip of a boa constrictor!

How can such a man be of much use in the service of the King of Kings when he has one foot in the temple and family of witchcraft?

A MASONIC "BAPTISM?"

Not only does Satan provide a counterfeit to the regeneration process, he also gives the Mason a baptism. As bizarre as it may sound, the Mason is forced to experience a parody of the rite of water baptism!

Most Christians understand that water baptism is a symbol of the new believer dying with Jesus by going into the water, then being raised with Jesus when drawn out of the water by the minister (Romans 6:4).

In a sacrilegious echo of that beautiful truth, the new Master Mason "dies" with the Masonic "christ," Hiram Abif, and then is "raised with him" with the strong grip of the Lion's paw by "King Solomon." This is the origin of the phrase, "being raised to the sublime degree of a Master Mason."

The resurrection with Hiram is intended to be what occultists call a "consciousness-raising experience." It is supposed to induce an altered state of consciousness, an open doorway for demon oppression!

Look at the ordeal to which the Master Mason is exposed. After going through an hour or two of the usual rigmarole he has come to expect from the first two degrees, he is dealt a humiliating psychological blow by the Lodge Master—he's told sternly that he is not a Master Mason, nor may he ever be![5] He is then warned that he may not even survive what he is about to experience.

He is blindfolded and led quickly around the Lodge, and violently accosted by three men. Finally, he is bashed on the head by a cushioned "setting maul" and caused to fall violently backward into a makeshift trampoline. He is then trundled about as a dead body, "buried" twice and finally "raised" by the strong grip of a Master Mason and given the secret Master's Word, "Mah-Hah-Bone," in a low breath by the Master, pretending to be Solomon.

This is a blatant recreation of the ancient pagan sex-cult resurrection rites mentioned earlier, and a mockery of the Lord's resurrection and Christian baptism! After "dying" with the Masonic "christ" (actually Ba'al or Tammuz), he is raised with him.

Not only has he been "Born again" with the umbilical cord of the witch goddess strangling his neck, he also gets a pagan baptismal rite! How it must grieve the Holy Spirit to see Christians submit to this profane rubbish!

VIRUSES IN THE BIO-COMPUTER

Then we have what Masonry does to your mind. I was "honored" to have as my Poster in the Lodge (the fellow who helps you memorize the ritual) an actual 33° Mason who was a Grand Lodge officer. He was a kindly, older gentleman who was patient with me as I struggled to learn the degree work necessary to advance.

When I got discouraged, he smiled and urged me on. He sat beside me in his living room and told me he had found the memory work did all sorts of interesting things to his mind. He was a big executive, and said the memory work clarified his mind and enabled him to have a more astute business sense. He credited it with his rapid advancement in a huge firm.

As I worked, I discovered my mind began to subtly change. I was beginning to think like a Mason. Running these strange, arcane sounding phrases over and over in my head had an almost hypnotic, mantra-like quality! I realize now that such phenomena were the first steps towards moving me in the direction of illuminism.

My mind was being given more "light." My "software" was being wiped clean and rewritten to conform to the mold of Masonry's dark god! In witchcraft, we called this "meta-programming" (meta meaning "beyond"). My brain tapes were being reprogrammed in the first step towards my supposed "evolution" from homo sapiens to what we called homo noeticus—the "new man," the super man!

Although this sounds strange, there is a lot of truth in it. Mentally, as well as physically, you are what you "eat!" If you spend your time reading the Bible, you conform to that way of thinking. Your "bio-computer" (as we called it) is programmed to think Godly, heaven-directed thoughts. This is why the Word of God commands us to read it avidly and think upon lovely and pure things. (Philippians 4:8)

But if all you read are trashy or occult books, it is "garbage in—garbage out!" Imagine if your poor brain was constantly filled with "tapes" of Masonry rituals playing over and over. Imagine those tapes as scientifically designed like spiritual computer viruses to get into the very depths of your mind's software and eat away everything worthwhile!

I was working on an assembly line when I was a first degree Mason, and it was mentally boring. I spent hours at a time fixing the Masonic memory work in my head. Although there is no way to prove this, I can look back and see that those countless hours I spent memorizing ritual work were times in which Satan had a chance to hot-wire my mind!

When I was Born Again, I realized that my mind was remarkably refreshed and regenerated by the blood of Jesus. Much of the Masonic mind patterns were instantly cleaned out. That is why, when I returned to the Lodge on that summer morning, I felt like such an alien, a stranger. I had to renounce my Lodge oaths and sever my membership.

MASONIC MUZAK?

It may seem I am carrying this computer metaphor too far. Can simple memory work in a Lodge really do spiritual damage? At the very least, it can cost the Christian Mason time which could have been spent in Bible reading, prayer, or Christian service.

How would you like to stand before Jesus and tell Him that you were too busy memorizing the twelve clauses of the Master Mason oath to go next door and tell your neighbor about Him? Not for me, thank you!

Beyond that, we have noticed a formidable spiritual deadness in talking with Masons about the Lord. It is like pouring water over a rock! What can this deadness come from?

When my wife and I were fairly new to the evangelism ministry, we travelled back to Milwaukee to talk to one of our former friends and pupils in witchcraft whom we had recently been blessed to help lead to the Lord. Both he and his younger brother had been in our covens, and he wanted us to meet with the brother to explain to him what we had discovered about Jesus.

I had led this younger brother, "Tom," to join my Lodge. He was one of the kindest, most tender-hearted and sincere people in any of our groups, and was deeply religious. However, he was not yet a Christian in the Biblical sense. He had followed us through Wicca, Masonry and Mormonism, and now we wanted to help him back out.

We spent hours with "Tom," and explained the gospel to him. He was willing to pray and ask Jesus into his heart. He was willing to prayerfully renounce Wicca and Mormonism. But in spite of his love and respect for us (which was mutual), he could not bring himself to see anything wrong with Masonry. When we spoke about it, his eyes would glaze over.

He was very proud of the memory work he had completed because he was raised in a disadvantaged home and didn't have a good grasp of the language. It was indeed an accomplishment to be proud of. However, he was so caught up in the work, he was willing to turn his back on Jesus. He could have let go of Mormonism and Wicca, but not Masonry.

We parted with hugs, but he did not pray to receive the Lord. Soon, in spite of his best intentions, he began to slide back into occultism. We haven't heard from him for several years, but we pray for him. His spirit, which was very sweet and searching, had been rendered inert by frequent exposure to the musty mummery of the Lodge! Satan had enshrouded his mind in the soft cotton batting of the ritual, insulating him from God's truth.

"Tom" is our dearest example of literally dozens of Masons we have witnessed to, only to find that Satan has cleverly quarantined their minds from our efforts. He has his "Masonic Muzak" tapes playing so loudly in their ears they cannot hear the gentle promptings of the Holy Spirit.

What a strategy! Not only has Satan succeeded in getting Christian Masons in spiritual bondage through blood oaths, sin and idolatry, he has even managed to capture their minds! He has turned the Lord's command on its head! Instead of conforming their minds and renewing them in Christ (Romans 12:2, Philippians 2:5, 4:2), he is turning them into little Masonic robots who can barely quote John 3:16, but can easily give you the entire G Lecture from second degree!

Satan has managed to mock not only the regeneration experience and baptism, but even the process of sanctification itself with his own twisted rituals. However, he is not content with entrapping the Christian Mason's mind, soul and body—he wants his family, too!

21

The Curse and the Hope

* A teenage girl has epileptic seizures so horrible that no medication can help. Her parents' renunciation of Masonry begins a process which ends with her being healed of the epilepsy by the power of Jesus!

* A high level officer in the Eastern Star wakes up to night terrors involving a goat-headed demon trying to molest her the night after she performed Star rituals. These awful occurrences stopped when she renounced the Star and quit.

* A grandchild with severe learning disabilities is miraculously brought nearly up to speed within days of his grandmother renouncing the Masonic ties in her family.

* A woman with crippling MS is brought into complete remission by cutting her ties to the Eastern Star and asking the Lord to forgive her for that association.

* A suicidal teen who is into satanism and black metal music comes forward for prayer and receives Jesus after her mom renounces the generational Masonic links through the child's grandfather.

These are just a few accounts of people (and often children) whose lives were severely hampered by involvement in Masonry. I mention them not to dismay you, but to show you there is no tragedy that the cross of Calvary cannot heal!

Satan can come in and wreak spiritual and physical carnage in a home when the father is a Mason! Parents who seem to be good, God-fearing people are bewildered that their child is ill, or their teen is involved with satanism, promiscuous sex, or trying to commit suicide. Because information such as you find in this book has been kept out of most Christian media, parents are often astonished to learn that Masonry can be such a cancer in the home.

This due to a lack of good teaching on the cult of Masonry in churches, and because of the lack of preaching on the issue of spiritual headship in the home. Many pastors are uncomfortable with the subject because they feel it is "chauvinist," or "un-liberated" to preach about the husband being the head of the wife (I Corinthians 11:3-10).

God's Word is never old-fashioned. For lack of sound teaching on this, many Christian marriages have shipwrecked, and many children have become lost in the raging seas of adolescence. It is essential that both the husband and wife understand the principle of headship in the home, and thereby comprehend the dangers of Masonic involvement.

HEADSHIP

Paul teaches that the "head of every man is Christ; and the head of the woman is the man" (I Corinthians 11:3). This is how men and women are created by God. Two other key verses in this passage tell us:

> But every woman that prayeth or prophe-
> sieth with her head uncovered dishonoureth
> her head: for that is even all one as if she
> were shaven...For the man is not of the
> woman; but the woman of the man. Neither
> was the man created for the woman; but the
> woman for the man. For this cause ought the
> woman to have power on her head because
> of the the angels.   I Corinthians 11:5, 8-10

Although this is a rather deep passage, it is not too deep that it cannot be interpreted by turning to the Bible itself for help. The first part is clearer. The husband is to be the head of the wife, even as Christ is the head of the husband. Paul clarifies this principle elsewhere, telling us:

> For the husband is the head of the wife, even
> as Christ is the head of the church: and he is
> the saviour of the body. Therefore as the
> church is subject unto Christ, so let the wives
> be to their own husbands in every thing.
> Husbands, love your wives, even as Christ
> also loved the church, and gave himself for
> it; That he might sanctify and cleanse it with
> the washing of water by the word.
>   Ephesians 5:23-26

It is not a question of the husband "lording it over" the wife as much as it is the husband giving himself up for his wife, laying down his life for her as Jesus did for the church. This is not a permit for abusive behavior on the part of husbands in any way, shape or form. The wife must be submitted to the husband, but the husband must be Christ-like in his care and devotion to the wife. He must be the channel through which the Holy Spirit can "sanctify and cleanse" his wife.

TO SPREAD YOUR COVERING

The last part of the passage quoted above is the most puzzling. What does "For this cause ought the woman to have power on her head because of the the angels" mean? This verse, and the ones preceding it, have often been interpreted with total literalness, to the extent that women have felt obliged to wear veils or hats to church.

However, the word here is "power," which obviously means more than a veil. If we stand aside and let the Bible interpret this passage, it all becomes clear. In Ruth 3:8-9, we find the elegant Hebrew custom to which Paul is referring:

> And it came to pass at midnight, that the man was afraid, and turned himself: and behold, a woman lay at his feet. And he said, Who art thou? And she answered, I am Ruth thine handmaid: spread therefore thy skirt over thine handmaid: for thou art a near kinsman.

Ruth is asking Boaz, her near kinsman, to marry her out of obligation to his dead relation, Mahlon, her former husband. For him to cover her with his skirt was a way of saying he would marry her, taking her under the mantle of his protection, even as we, the church, are under the mantle of Jesus' protection.

When a Godly man marries a woman, he takes

"power" over her because of the angels. Which angels are being referred to is unclear. Some say they are not good angels, but fallen angels, which might attack or tempt the wife. Others say they are good angels who are scandalized by seeing a wife out from under the authority of her husband.

Although we cannot be certain, I prefer the first explanation, especially since there are two other places in scripture which seem to indicate that fallen angels might be very dangerous to unprotected women (Genesis 6:2, Jude 6-7). Thus, it is the duty of a man to provide a spiritual covering for his wife.

Why is this? Because it is the way the Lord made the marriage relationship. It is the way men and women are put together. In a good, Christian marriage, the husband is the covering for the wife— her shepherd, her "lightning rod," if you will. He takes all the flak for her, even as Jesus did for us. He should be both a unique font of blessing for her and her strong protector from attack, whether spiritual or physical.

A Godly woman can maintain herself in marriage to a backslidden or unsaved husband, but it is a struggle. Many times, such men are saved or brought to repentance by the patient witness of their wives (I Corinthians 7:14). In such situations, the woman often becomes the protector of the weak man. This is not what God intended, but it does happen in a fallen world.

This is why Freemasonry is such a wretched spot upon the feast of marriage. Spiritual headship is a two-edged sword that can cut both ways. If a husband is in deep sin, like Masonry, then all the evil spiritual power of the Lodge seeps through him into his wife and (worse yet) into his children, even if they never set foot inside of a Lodge building!

So here comes "hubby" home from the Lodge meeting, trailing clouds of Masonic imps in his wake as he walks in the door. It is just like bringing home the flu bug from the office, except this "bug" is spiritually more contagious, and it takes more than fluids and bed rest to kick it!

THE DEVIL TAKES THE WEAKEST

If, as mentioned above, the wife is a really strong personality and solid in her relationship with the Lord, she may be able to weather the buffetings of her husband's idolatry.

However, the children are another matter altogether. They cannot be expected to fend for themselves spiritually, any more than they can physically. Tragically, the children bear the brunt of their daddy's Masonry.

This is not just because of the likelihood of the kids getting involved in DeMolay, Rainbow or Job's Daughters (Masonic youth orders), although that danger is real. More important, they will be spiritually affected by the idolatry in their father's life.

Even though there might not be any problems evident at the moment, parents need to be aware of the long-range dangers of their involvement in the Lodge. This places a lot of unseen stress upon the children which might not actually show up until periods of crisis in the child's life. Frequently, one of the more serious developmental crises is the onset of puberty! This is often when literally "all hell" often breaks loose.

It might be compared to a hairline crack in an airliner. It is invisible to the naked eye and seems to cause no problems. However, Masonic involvement is like a deathwatch beetle in the family which eats at it spirituality from the inside out.

Like that hairline crack, it never shows until some extraordinary stress hits the person, a spiritual "wind shear" to continue the metaphor—a crisis in the family or a personal crisis—and the crack widens into a gaping maw and an engine falls off. We then have a spiritual "crash and burn."

The Word of God speaks directly to this awful consequence:

> He that troubleth his own house shall inherit
> the wind: and the fool shall be servant to the
> wise of heart.   Proverbs 11:29

Quite often, during puberty, when the young person is struggling with key issues like body image, sex drive, and identity formation, the hidden spiritual weaknesses caused by the father's (or mother's) involvement in Freemasonry causes the crack to blow wide open.

FUEL ON THE FIRES

Two of the most powerful forces working on the teen are the sex drive and the need to rebel and assert his or her own individuality. Both are part of the craziness of being an adolescent, and are made incredibly worse by Masonic headship in the home. Why is this?

As made abundantly clear earlier, Freemasonry is essentially a fertility cult—a cult revolving around sexual reproduction. Its sacred symbols are sexual idols. In magic, these idols are especially designed to invoke strong sexual urges and fertility. In simple terms, they are talismans or magical devices to increase lust! Obviously, the last thing a teenager needs is more lust!

He or she already has a body full of raging hormones, and is surrounded by a culture full of sexual images and pornographic music and videos. To have a father bring home spiritual influences from the Lodge which increase lust is adding gasoline to an already spreading blaze! The lewd influences of Masonic principalities are amazingly strong. Even many adults cannot resist them, and it's surely asking too much of our immature teens.

How can we expect our children to wait until marriage when our involvement in an ancient mystery religion has opened the floodgates of desire into our homes?

How can we parents expect our children to respect authority, and honor the Word of God and its teachings when we refuse to obey it? Children have acute noses for hypocrisy, and if they know their Bibles at all, they will easily see that Dad should not be shuffling off to the Lodge meeting every Monday night. His actions will speak louder than his words.

Additionally, although most kids will probably not know very much about the Masonry their father and/or mother is involved in, the fact that their parents are members of "the world's largest coven" brings the sin of witchcraft into the home. The Bible teaches us that:

> ...rebellion is as the sin of witchcraft, and stubbornness is as iniquity and idolatry.
> I Samuel 15:23

Parents wonder why their kids are rebellious, and yet they are out of God's will by being in their Lodge. The parents' rebellion will simply percolate down to the children, and especially during the teen years, may erupt into an explosion of anarchy!

THE SINS OF THE FATHERS...

There is a TV spot of a parent challenging his teen about his drug use, only to find that his son learned about drugs from watching the parents. The announcer ominously intones, "Parents who use drugs have kids who use drugs." We could just as easily say, "Parents who do idolatry have kids who do idolatry."

Many commentators have observed the dramatic increase in the following practices among teens in the past couple of generations:

1) Pre-marital sex

2) Abortion

3) Illegal drug use

4) Dabbling in the occult and satanism

It would be simplistic to blame any one thing for all these phenomena. They are interrelated. Consider the fact that all these adolescent sins (and they are sins) are directly tied into Freemasonry.

1) Pre-marital sex

As already mentioned, Masonry is a cult which promotes carnal appetites by its spiritual influence. No wonder the children of Christians began to experiment with illicit sex when membership of their parents rose in Masonry in the sixties.

2) Abortion

This is child sacrifice. It has been "sterilized" and clothed in medical garb, but it is literally a mother (of whatever age) placing her baby on a pagan altar to be killed. It doesn't matter whether the sacrificing high priest is a black-robed satanist with a dagger, or a doctor in white with his suction machine.

Our aborted children are being sacrificed on abortion altars to the "gods" of convenience, "reproductive freedom" and the "right to privacy."

What does this have to do with Freemasonry? The god of the Lodge is Ba'al, or Abaddon, or any number of other pagan gods. One of the most prominent of those gods in the Bible is Molech. Both Molech and Ba'al demanded infant sacrifices from their devotees. The Bible calls this detestable practice, "giving your seed to Molech" (Leviticus 20:2) or passing "your seed through the fire to Molech" (Leviticus 18:21). It was a capital offense! This is how ancient mystery cults worshipped in many cultures according to the Bible and secular historians.[1]

If the father of a girl has knelt and sworn allegiance to a god like Ba'al or Molech (however unknowingly), should we be surprised that his daughter finds it easy, or even necessary, to offer up her baby in accord with that god's evil liturgy? This is something to consider in the midst of our Godly crusade for the rights of the unborn.

Are we fighting all the right battles in the courts and the media and the streets in front of abortion clinics, and then losing the war because we have opened our homes and families to the strongman of Molech by joining his Lodge?

Perhaps if Masonry were cast out of our churches and homes, our daughters would not find it so easy to consider abortion. We must wrestle first and foremost with these spiritual strongholds,and Freemasonry is one of the biggest (Ephesians 6:12)!

3) Illegal drug use

Witchcraft and drugs go together like the proverbial horse and carriage—you can't have one without the other. The first drug peddlers were witches, and a major part of the ancient mystery cults was the ingestion of hallucinogenic drugs.[2]

We should not be surprised that children of those involved in Masonry are susceptible to drugs. The "doorway" has been opened by their folks' involvement in the Lodge, even if their folks never touched a drug themselves. Remember that kids, and even adolescents, are much weaker spiritually than their parents, and cravings which the parents can resist are all too compelling for a youth.

4) Dabbling in the occult and satanism

If a child's parents are "unequally yoked" (II Corinthians 6:14) in fraternal fellowship to a worldwide organization with thousands of witches and satanists, that child will be open to the siren song of the occult.

Through Freemasonry, the tentacles of witchcraft reach deeply into thousands of Christian men and their families. It is the Trojan horse of contemporary American Protestantism which has rendered many of the mainline churches stone-cold dead!

STOPPING THE JUGGERNAUT!

Satan is too wily a strategist to put all his serpents' eggs into one basket. He launched a multi-taloned assault against our young people—TV, rock music, secularism in schools, drugs, etc. But the first step he had to achieve was fouling the nest of the Christian church by bringing Freemasonry into it.

Even though Masonic Lodge activity is on the wane in most areas, the vile spiritual momentum created by the parents and grandparents' involvement in Freemasonry has set in motion a snowball of sin.

Like an unstoppable juggernaut, this current of pagan mystery religion draws into itself all the side streams of occultism, MTV, drugs and slasher videos until it lumbers on under its own power, crushing an entire generation of children under its malignant weight!

Only the power of the Cross can stop it! Nothing but the Blood of Jesus can bring the juggernaut of corruption to a halt! But with it must come true brokenness and heartfelt repentance! With it must come a willingness to renounce Masonry for the evil that it is, in spite of all its cosmetic niceness.

With it must come a genuine desire to "come out from among them and be ye separate," no matter what friends, family or business associates might say.

If the Christian Mason loves his family, and more importantly, if he really loves Jesus Christ, he will be willing to lay aside the "unfruitful works of darkness" (Ephesians 5:11) no matter how glamorous they might be. He will be willing to call sin by its name.

Only by doing this, and by following the suggested procedure in the concluding chapter of renouncing the practices of Masonry before the Lord and legally resigning from his Masonic memberships, will he be able to stop the juggernaut of evil.

Jesus can break the chains of sin and death which bind the Christian Mason in the Lodge, but he must repent and confess to the Lord. Only then can the momentum which threatens him, his family and his church be brought to a grinding and impotent halt.

Sadly, the Christian Mason will need to count "the cost" (Luke 14:28). I would be less than honest if I did not mention that the wrath of the Lodge can be massive and frightening. Careers have been ruined, reputations destroyed, and even homes or businesses burned to the ground by Masons angered because their "toy" has been criticized. Even death threats occur.

This shows how deep the wellsprings of the Deceiver have been drilled into the hearts of Freemasonry's servants.

These violent acts are rare, but they can occur. It would not be the first time a Believer in Jesus suffered persecution or martyrdom for his testimony. A true bond slave of Jesus Christ should count all Masonry "as dung" (Philippians 3:8) compared to the boundless joy of serving Him!

> ...all that will live godly in Christ Jesus shall suffer persecution   II Timothy 3:12

The newly set free Mason should seek prayer support and discipleship from men of God with true shepherds' hearts! He should look to Jesus, but also be aware of the harm the Lodge can work on those

240

he loves most. He should, by the power and authority of Jesus, trample the demonic playthings of the Lodge beneath his feet and "forgetting those things which are behind, and reaching forth unto those things which are before," he should press toward the mark for the prize of the high calling of God in Christ Jesus (Philippians 3:13-14).

22

How To Protect Your Family

Throughout this book, you've seen a grim picture of the havoc that Masonry has created within the Christian home and church.

You've seen how the involvement of either parent, especially the father, in the Ba'al worship of the Lodge can cause strongholds of evil and rebellion in the home, in spite of the best efforts of the parents to maintain Christian standards.

Jesus Christ offers a simple solution to these problems through His Cross and shed blood. Nevertheless, Christians involved in Masonry must be willing to reach out to Jesus. They must confess Masonry as sin, renounce it and get it out of their lives and out of the lives of their families. Only then can they confidently resist the activity of Satan against them and those most dear to them.

The vast majority of Masons have no idea of the

danger to which they have been exposed. They are victims. They do not know that the symbols of the Lodge which they wear so proudly are idols of lust. They have been deliberately deceived by those in the highest reaches of the Craft. However, that does not excuse them. Hosea 4:6 still applies.

If you drink a soda with poison in it, it will kill you, even if you didn't know the poison was in it. If a baby, ignorant of the law of gravity, crawls off a sofa, it will fall with a jarring thud to the floor. These realities, which everyone accepts, are based upon laws created by the Lord. Ignorance of these laws does not protect one from the consequences of their violation.

Ignorance of the spiritual venom of Masonry does not protect the Mason or his family from the judgements of God, or from the buffetings of Satan. And yet, people who accept the laws of gravity as solid and inviolable neglect to realize that the same God who designed those laws also designed spiritual laws which are just as powerful and just as difficult to circumvent.

The law of God with regards to sin (any sin) in the life of a Christian is simple. The apostle John says:

> This then is the message which we have heard of him, and declare unto you, that God is light, and in him is no darkness at all. If we say that we have fellowship with him, and walk in darkness, we lie, and do not the truth: But if we walk in the light, as he is in the light, we have fellowship one with another, and the blood of Jesus Christ his Son cleanseth us from all sin. If we say that we

243

have no sin, we deceive ourselves, and the truth is not in us. If we confess our sins, he is faithful and just to forgive us our sins, and to cleanse us from all unrighteousness.

1 John 1:5-9

Masonry, just as any other sin, needs to be confessed. There must be repentance, and opportunities for the return to the sin need to be shut off. That means the Mason who is a Christian must formally resign from the Masonic order and all associated organizations by written letter.

DO I HAVE TO RESIGN?

Some question why this step is necessary, and ask, "Isn't it enough to go to God in repentance? Why do I need to formally resign?" This is partially a heart matter, but there's also the issue of giving the devil no place in our lives (Ephesians 4:27).

If a Christian was struggling with alcoholism, we would have trouble thinking his repentance was genuine if he kept bottles of whiskey stashed around the house. If a Christian had a problem with pornography and had repented of it, we might doubt his sincerity or faith if we discovered he still has smutty magazines in his bedroom. It's like saying in their hearts:

I am asking the Lord to forgive me, and to help me stay away from this sin, but just in case He can't, I'm hanging on to a couple of Playboy magazines to tide me over.

It shows a lack of true faith in the redemptive power of Jesus to truly set one free. The Bible's counsel is to destroy all Masonic books or artifacts (aprons, rings, fezzes, etc.) or, in the case of books, perhaps turn them over to reputable ministries for research. (See Acts 19:18-20.)

Satan is the accuser of the brethren (Revelation 12:10). If you hold legal membership in the Lodge, even if you have not been inside a Masonic temple in years, he can use that to attack you, and destroy your testimony. He can whisper:

> You aren't really set free from Masonry. You're still on the books. What kind of Christian are you?

He can also try and get you to return to the Lodge, if only to see old friends.

More significantly, Satan can use that legal connection between you and the Lodge to attack your family. When you, as an ex-Mason, pray for the protection of the blood of Jesus over your wife and children, Satan can stand before God's throne accusing you. He can truthfully say:

> This man has not really repented. I still have the right to trouble his house because he still is legally connected to my Lodge. He is still serving me.

That's why the sincere prayers of many Christians bounce off the ceiling. Satan is the ultimate legalist, and he has a mind like a steel trap! Think of him as the world's most persistent prosecuting attorney. If he can find the tiniest loophole through which he can attack a Christian, he will. Retaining your Lodge membership provides him with just such a loophole.

THE INIQUITY OF THE FATHERS

Freemasonry is not just any sin. It is the sin of idolatry! It is the worship, knowingly or unknowingly, of false gods—gods which have been

245

worshipped by witches and pagans for thousands of years. Because of the special character of this sin, it merits an extraordinary clause in the Ten Commandments. The Lord commanded:

> Thou shalt not make unto thee any graven image, or any likeness of anything that is in heaven above, or that is in the earth beneath, or that is in the water under the earth: Thou shalt not bow down thyself to them, nor serve them: for I the LORD thy God am a jealous God, visiting the iniquity of the fathers upon the children even unto the third and fourth generation of them that hate me.
>
> Exodus 20:4-5

Notice that idolatry visits the sin of the fathers to the children, and even to the great-grandchildren! What a terrible trap the Deceiver has laid for the Mason. Not only is he in spiritual trouble, but unless he repents and asks for the intervention of the blood of Jesus, his posterity will also be oppressed!

People have often come to us for counsel unable to understand why they are having such a problem walking with the Lord. Upon questioning, it frequently emerges that there's a father or grandfather who was in Masonry. Scarcely a family in the United States is free of this wretched plague for four generations back!

Another generational curse that may beset believers is the ten-generation curse of Deuteronomy 23:2. God commands that a person of illegitimate birth must be kept out of the congregation of the Lord, and his or her ancestors for ten generations! This does not affect a person's salvation, but it can prevent a victory in their walk with the Lord.

Very few people can trace their family line back ten generations, and human nature being what it is, somewhere in those ten generations there may have been an ancestor born outside of wedlock. It cannot hurt to ask the Lord to cleanse the family line of that anathema. A brief prayer suffices!

It is exciting to see the relief and peace on their faces when they simply go to their sweet Lord Jesus in prayer and ask Him to forgive and cleanse those generational curses. In minutes, they have new joy in their hearts! A huge door has just been slammed in the face of Satan, and he can no longer bother them and their family!

Besetting sins and health problems often begin to disappear with this confession. It is truly amazing to see Satan quickly flee like a scalded cat before the blood of Jesus when he no longer has a legal right to remain.

Some people may find it odd to confess the sins of their ancestors. Is this Biblical? Psalm 106:6-7, Nehemiah 9:2-3 and Daniel 9:5-6 each contain mighty prayers of confession in which the sins of the fathers are repented of as well. Obviously Nehemiah and Daniel took this commandment very seriously.

LAW OR GRACE?

There may be a protest that:

> This is Old Testament stuff. We aren't under the law, we are under grace.

For the Christian, this is true in a spiritual sense. None of this affects the person's salvation. Those in bondage to Masonry or under a generational curse of Masonry will go to heaven when they die. Their spirits have been regenerated!

247

However, Satan will do his level best to make their trip through life miserable. Some of the more "triumphalist" Christians forget that we live with a foot in the Kingdom of heaven and a foot in the fallen world, where the Law still reigns over our soulish and physical bodies.

Otherwise, Christians would not die or get sick. Christian women would not have labor pains. Christian farmers would have no thorns or thistles to contend with.

Although the Christian is seated with Jesus in heavenly places (Ephesians 2:6) during this life, his body and soul are all too earthbound. The Law can still have place in our mortal members, especially if there is unconfessed sin! Most attacks against the families of Masons are of a physical or soulish (emotional, mental, or volitional) nature.

Sickness, rebellious children, mental disorders, and drugs all bedevil the family with Masonry in its veins. Whether Masonic involvement can ultimately cost the Mason his eternal life is too terrible a question to even contemplate. I would prefer not to think so. However, he is definitely tempting the wrath of God by kneeling at the altars of Ba'al and frittering away his time in the service of some other Worshipful Master.

The tenth chapter of Hebrews contains chilling warnings for those who think they can tempt the mercy of God. As verse 31 says:

It is a fearful thing to fall into the hands of the living God.

If this doesn't make sense to you doctrinally, I encourage you to check it out in the Word and avoid all pre-conceived notions (Acts 17:11). In years of praying with Masonic families for deliverance from these curses, we have seen wonderful testimonies of the power of Jesus to save to the uttermost!

The Biblical principles are ignored at the peril of the Masonically involved family! Take the Lord at His word, even if it does rattle your theological cage a bit. You have nothing to lose but your squares and compasses!

DOWN TO BUSINESS

Having laid the groundwork, let us consider the power of the cross and blood of Christ that can be brought to bear on the weak and beggarly elements of the Lodge.

If you are a Mason who isn't sure if you are saved by the blood of Jesus, see the end of this chapter for your instructions.

If you are a Mason who is a Born Again Christian, here is what you need to do. This is a quick checklist. Examples of prayers for these will appear in Appendix I.

First, ask the Lord to forgive you for the sin of Masonry. You need to renounce the Lodge before Him in prayer as a false religious system and ask the help of the Holy Spirit to never return to it.

Second, ask God to cut all generational ties between you and any Masonic ancestors, and any ties between you and your own children or grand-children (if any), and cleanse all those ties in the blood of Jesus Christ. We have found that it helps to seal these prayers with a simple anointing with oil upon the forehead of the person renouncing these things and/or family members, as a token of the reconsecration or "resealing" of your body, soul and spirit as cleansed Temples of the Holy Spirit and yourself as a servant of the living and true God (James 5:13-14, Revelation 7:3).

Third, destroy or dispose of any Masonic regalia you have in your home. Aprons, fezzes, books and anything combustible should be burned. Jewelry should be destroyed or, if extremely valuable, melted down and recast. We'll cover this later.

Fourth, write a letter to your Blue Lodge and any other Masonic bodies of which you are a member, asking that your name be removed at once from the membership rolls. The letter should be charitable and brief, but it would not hurt to share the reasons you are doing this. A sample resignation letter is included in Appendix II.

Fifth, if your involvement in Masonry has been conspicuous or highly public (i.e., if you were a Lodge master or a Grand Lodge officer), which may bring scandal to the cause of Christ, it may be necessary for you to make a public statement of repentance in your church. Seek the guidance of your pastor on this (unless, of course, your pastor is a Mason himself)!

If your wife or children have been active in Masonic organizations like the Order of the Eastern Star, DeMolay, or Job's Daughters, they need to go through similar steps themselves. This brings up another area which may need to be dealt with.

THE SOUL TIE

"Soul ties" are links formed by close associations with people we love, especially in sexual intimacy.

God ordained the marriage covenant to be the coming together in "one flesh." Originally, there was no concept of a priest or rabbi blessing a marriage. The very act of sexual relations constituted a marriage. The Bible tells us this is how the early patriarchs took their wives (Genesis 24:67; 29:23).

This means that intimacy with a person creates a covenant link. They become one flesh. Thereafter, whatever happens to the one partner can percolate into the other partner. It may not always happen, but it can. If a husband is involved with Masonry, the bondage he experiences in the Lodge is probably seeping over into the wife and robbing her of much spiritual victory.

Because of the principle of headship, a husband is supposed to be a spiritual protection. However, by allowing Masonry to foul his spiritual "nest," he is bringing curses upon the home instead of blessings.

If the husband, through Masonic involvement, has allowed demonic access into his soul, those demons can come through him into the wife unless she breaks those ungodly soul ties through prayer.

At the risk of being grotesque, bondage and demonic entrance are the ultimate forms of venereal disease, spiritual AIDS, if you will. Until the husband (or wife, if she is the one Masonically involved) renounces the Lodge completely, it is possible that marital intimacy can quickly re-open doorways of bondage or demonic access.

Sadly, some forms of Masonry involve at least the recreational (if not ritualized) use of prostitutes. Shriners are especially noted for this vile practice.

Both husbands and wives should realize that the reason the Lord guards and protects marital chastity so carefully is that involvement with any adulterous relationship brings bondage with it. The unfaithful spouse can bring demonic spirits home with him from his Shrine conventions and afflict his more spiritually sensitive wife with unimaginable torment when they come together again in the marriage bed (I Corinthians 6:15-17).

Marital infidelity not only brings with it the danger of physical disease, but of spiritual darkness as well. We have had to pray with several wives of Shriners for deliverance from demon oppression, in spite of the fact that these women were doing everything they could to live for the Lord Jesus.

This is tragic, especially since it can be stopped by repentance on the part of the sinning party, and by cleansing the covenant ties between spouses with the blood of Jesus.

WHAT ABOUT ALL THAT JEWELRY?

When we discuss the destruction of Masonic regalia, cries immediately come up from some quarters. Masonic jewelry can be quite valuable, as Freemasonry often attracts the rich and powerful. Large numbers of diamonds and/or precious stones, and sizeable gold rings are often among the trinkets which need to be destroyed.

The ideal would be total destruction. This would ensure that the spiritual evil upon them would not be passed on to some other unsuspecting person. I believe that God would honor such a sacrifice of faithfulness and obedience in wonderful ways.

I personally had more than $3800 worth of

esoteric books and ritual regalia when I got saved. Within a few weeks, the Holy Spirit gave me no peace until I disposed of them. After some time on the phone, I found a pastor who was willing to help me burn or destroy all of it!

The wonderful thing is, three years later, I got into evangelism ministry to the cults. In the writing and research I was doing, I needed some of the rare material which I had destroyed. I prayed and gave it to the Lord. Within a month, my wife and I led a former fellow witch to the Lord, and within a few weeks he led another magician to Jesus.

This caused a chain reaction, and he showed up at my door one day before we left for full-time ministry with a trunkful of esoteric books from each of them. Praise God for His faithfulness! The Lord answered my prayer and restored about 90% of what I had burned, now that I was in full-time ministry and mature enough to use it.

I tell this story only to show how God is pleased to bless those who aren't afraid to go out on a limb in faith for His service.

The wholesale destruction of jewelry may be more than some Masonic families (or pocketbooks) could stand, and there may be another option. This involves either melting the jewelry, or removing all Masonic markings from it. Some people have had a jeweler refinish or recast their Masonic rings, perhaps with a Christian symbol. This may be an option, especially if the ring is costly or has precious stones in it.

Biblically, this would be "spoiling the Egyptians" (Exodus 3:22). The Israelites took much jewelry from the Egyptians when they left Goshen, and much of this was probably idolatrous, considering the religion of Egypt. The vast majority of the gold was melted down or beaten into the very components of the ark of the covenant, the candlestick and other sacred things for the tabernacle in the wilderness.

If you choose to rework the jewelry, simply pray over the ring, anoint it with oil and bless it after it has been refinished, and dedicate it to the Lord, as was done in Israel.

If that is not a viable or practical option, and the only alternative is to sell it, that does accomplish your number one objective, which is to get the stuff out of the house. It is not the best arrangement, but it would still be God-honoring. I would pray over the materials and plead the blood of Jesus over them, and bind away all Masonic spirits before I turned it over to the new owners.

You might consider giving the proceeds from the sale to some worthy Christian mission organization, and certainly you should tithe out of the proceeds.

Even as I am concluding this chapter, I received a wonderful report from a mother who came to us for counseling. Her son, who is a Christian, was having trouble lately in His Christian walk. He also felt ill at ease in his bedroom. His grandfather had been a Shriner, and upon his death, had bequeathed his fez to his grandson, who had been building a formidable hat collection. The boy valued the fez as a treasured legacy from his grandpa, as well as for its exotic beauty.

His mom brought him to our office. Upon finding out about the evil nature of the fez, his response was truly noble! When he learned it should be destroyed, he asked if he could do it himself. We took it out on the back steps, and with a hammer he smashed all the jewels on it, including the Islamic crescent and scimitar, then gave it to us to be burnt.

A month or two later, the mother informed us that her son now has incredible peace in his room, and his walk with Jesus has taken off in incredible ways! God blessed the sincere sacrifice of that young man, and He will bless yours as well.

These are genuine spiritual principles, and we ignore them at our peril, and the peril of those we love!

23

Taking the Battle to the Enemy

Although we must first cleanse our homes of the evil of Masonry, for those concerned about prayer for your community and nation, there is another battlefront we need to discuss.

As most believers know, one of the very first obligations of any Christian is to pray for their rulers (I Timothy 2:1-2).

Another issue emerges as we go into intercessory prayer for our leaders. Many lawyers, judges and politicians are Freemasons. Some of the most powerful senators, congressmen and supreme court justices are Masons.

Additionally, most of our public buildings, both in Washington, DC, and in many state capitols, have cornerstones which were "dedicated" by Masons. This amounts to a luciferian curse!

Many Christians are bewildered by the "double-signals" they receive from politicians and even from

Christian teachers. Often, we are told that such-and-such a politician is a "great man of God" or a "solid Christian," yet their voting record is curiously off. Allowing for the evident need in politics for compromise, it still seems as though two agendas are operating in many instances.

Similarly, although many Christian leaders talked about America being a "Christian nation," and about our Founding Fathers' faith in God, some of this is misguided. Many of our nation's founders and early heroes were Freemasons, including Jefferson, Franklin, Washington, John Paul Jones, Paul Revere and (oddly enough) Benedict Arnold.

Some of those, like Franklin and Jefferson, were also Deists and/or occultists. Franklin belonged to the infamous "Hellfire Club" in London, a satanic society. Jefferson was said to be a Rosicrucian. Though a brilliant man, Jefferson snipped out passages from his Bible until he created the celebrated *Jefferson Bible* which removed all references to sin, atonement and the divinity of Jesus Christ. Such a man could hardly be called a Christian!

It is interesting to note, however, that for all the PR the Masons try to wring out of George Washington's involvement with the Lodge, it appears that actually his commitment was quite minimal. Although countless Masonic pamphlets have the picture of Washington presiding over a Grand Lodge in full Grand Officer regalia, that picture has about as much truth to it as the story of little George and the cherry tree!

Washington, who by all accounts was a serious

Christian, actually had no patience with the Lodge, and seemed to have scarcely been involved in it at all after joining it in 1752.[1] In fact, Washington wrote a Rev. G. W. Snyder, in 1798 that far from "Presiding over the English Lodges in this country, Washington" had not even been inside a Masonic Lodge, "more than once or twice within the last thirty years."[2] Washington's involvement with the Lodge has been vastly overstated by Masons eager to trade upon his reputation.

ESAU AND JACOB

When the New World was discovered, Europeans came to the colonies with two different visions for this new land. The Puritans and others came seeking religious freedom and envisioned a chance to establish a Bible-based civilization.

But others saw America as the "New Atlantis," a place where occultism could prosper uninhibited by Christianity. Long before 1776, a colony of Rosicrucians was established in Ephrata, PA! Colonial America was full of occult groups, and both Witchcraft and Freemasonry came over on the boats quite early!

Almost as in the birth of Esau and Jacob, from the birth of America there has been a struggle between two forces, even "in the womb." America was born out of uneasy compromise between Christianity on one hand and Masonic occultism on the other. This tension still exists today.

Many "Christian" politicians, including popular conservative senators like Jesse Helms, Strom Thurmond and Robert Dole are 33° Masons, and many lesser known current or past politicians like

Senators Nunn, McClure, Stennis, Hatfield, Johnston, Burdick, Glenn, Hollings, Bentsen, Stafford Grassley, Specter and Simpson, and Congressmen Jim Wright, Don Edwards, Claude Pepper, Dan Glickman, William Ford and Trent Lott are Masons![3]

Former president Gerald Ford was 33°, as was Gen. Douglas MacArthur. Other presidents who were Masons include Buchanan, Garfield, Harding, Jackson, McKinley, Monroe, Polk, Teddy Roosevelt, Franklin Roosevelt, Taft, and Truman.[4] Davy Crockett and Sam Houston were also Masons.

When I was a Mason (1975-1984), it was common knowledge among all my Lodge brothers that George Bush held the 33°. Yet his office now sends out a letter saying that the president "is not" a Mason. It does not say that he never was a Mason, and he may have demitted (left the Lodge) just to avoid alienating the Religious Right.

Christian Celebrities like Roy Rogers, Burl Ives and Norman Vincent Peale are also 33°! Ronald Reagan, long the darling of the Religious Right was made a Mason while in the White House. He was later implicated in the practice of astrology and carries a lucky rabbit's foot!

We do not deny the sincerity of these men's faith, but they are either deceived or have chosen to make a "devil's bargain" to reach the corridors of power. They may have realized that few politicians can attain prominence today without bowing the knee to Ba'al in the Masonic Lodge. This is why we desperately need to pray for our leaders!

It is also vital to counterbalance this list of "heroes," both present and past who have been

Masons with men of the Christian faith and U.S. history who have gone on record condemning the Lodge. These include Charles Finney (a former Mason), Dwight Moody, R.A. Torrey, Billy Sunday, John R. Rice (a former Mason), John Quincy Adams and Ulysses S. Grant,

A CHANNEL FOR "THE CHRIST?"

One of the most recent manifestations of this spiritual struggle is the all too evident alliance between the New Age movement and Freemasons.

Most Christians understand something of the New Age, that it is a blend of Hinduism, psychology and Gnosticism. Though the New Age concepts have been around since the Garden of Eden, the current manifestation of these beliefs can be traced back to the Theosophical Society, and its offshoots, Alice Bailey and Lucis Trust.

The New Age movement has become a uniquely American blend of Hinduism and "yuppiefied" spiritism. One of its key doctrines is the coming of a christ, referred to by most New Age leaders as Lord Maitreya. Although we cannot get into an extensive analysis of the New Age here, it is important to understand that New Age teaches the following:

1) God is an impersonal force, not a personal Being.

2) All are destined to become gods through spiritual evolution.

3) All are children of God. There is no need to be born again.

4) "The Christ" is not Jesus, but Jesus' teacher, Maitreya.

5) There is no real death, only reincarnation.

6) There is no sin, only the Hindu concept of karma.

7) All religions lead to God. There is no one, true path to Him.

8) Religions which teach there is only one true path to God (i.e., Judeo-Christianity) are anti-evolutionary and will need to change their thinking or be destroyed.

Obviously, this is contrary to Biblical truth, but very akin to Masonry's generic theology. A key part of the struggle for the souls of men in this nation is taking place on subterranean levels through the dove-tailing of the New Age with Masonry.

In addition, members of the Masonic Lodge and the New Age are infiltrating Christian churches.

Today leaders of the New Age movement are saying that Masonry will be both the political and spiritual channel through which the "Christ" will manifest. A letter from the Tara Center, a major New Age group states:

> The Masonic movement is one of the three main channels through which the preparation for the new age is going on. In it are found disciples of the Great Ones who are steadily gathering momentum and will before long enter upon their designated task.[5]

The same letter quotes Alice Bailey, an early New Age leader as saying:

> The Masonic movement...will meet the need of those who can, and should wield power. It is the custodian of the law; it is the home of the Mysteries and seat of Initiation. It holds

in its symbolism the ritual of Deity, and the way of salvation is pictorially preserved in its work. The methods of Deity are demonstrated in its temples, and under the All-seeing Eye the work can go forward. It is a far more occult organization than can be realized, and is intended to be the training school for the coming advanced occultists.[6]

We see that through the Lodges, the New Age leaders hope to influence not only individuals, but the entire global consciousness!

Benjamin Creme, the "prophet" for this Maitreya, and one of the most famous New Age leaders in the world, has written:

Through the Masonic tradition and certain esoteric groups, will come the process of initiation. In this coming age millions of people will take the first and second initiation through these transformed and purified institutions... The new religions will manifest, for instance, through organizations like Freemasonry. In Freemasonry is embedded the core of the secret heart of the occult mysteries—wrapped up in number, metaphor and symbol. When these are purified...these will be seen to be a true occult heritage. Through the Orders of Masonry, the Initiatory Path will be trodden and Initiation will be taken....[7]

There are millions of Masons in the world. Though only the tiniest fraction of those men understand the real spiritual nature of the Lodge, every one of them has knelt at the altar and been "plugged in" to the Masonic current—the "light!"

Through their "raising," what we have called

the Masonic baptism), they have been engrafted into Lucifer. The strains of Masonry have been imbedded deep in their psyches. It may lie there dormant for decades in 99% of the Masons, like a computer virus waiting to be activated. But when this false christ wishes, it will be suddenly energized.

In a spiritual sense, every Mason is like a time bomb waiting to go off. They do not realize it, but ticking deep within them is a satanic timepiece with an evil agenda.

When this false messiah appears, he will punch a psychic button and Masons will rush to obey! From the pinnacle of the Masonic pyramid, currents of malignant power will pour down through the ranks and every Mason will feel the perfidious purposes of "Lord" Maitreya quickened within him.

If this sounds preposterous, then consider how, in many churches, Satan has already called in his favors. I have spoken in churches where the pastor was brought down by Masons because he dared to stand for Biblical truth!

I have been in churches where Masonic members loved Masonry more than Jesus, where they disobeyed their pastors to follow their Worshipful Master.

These men are totally devoted slaves to the Lodge! They will march forth as pawns to do their new master's bidding, even if he is a false christ. Imagine if such a "pawn" were the President of the United States. At least 17 presidents have been Masons, including most within recent memory.

This is war, and it must be fought with spiritual weapons! We dare not trust these Masons and their claims of Christianity. They may be Christians, but they are dual loyalists, whether they know it or not. A dark piper lurks within them, waiting patiently to play a tune to drown out whatever is left of the Holy Spirit. We must pray for our leaders, especially if we know them to be Masonic "Christians!"

STANDING IN THE GAP

A sobering word for those who intercede is the passage in Ezekiel 22:30-31:

> And I sought for a man among them, that should make up the hedge, and stand in the gap before me for the land, that I should not destroy it: but I found none. Therefore have I poured out mine indignation upon them; I have consumed them with the fire of my wrath: their own way have I recompensed upon their heads, saith the LORD God.

The image here is almost like the tale of the little Dutch boy with his finger in the dike! God is looking for those who will stand in prayer in the breaches, and will spiritually plug the holes in the walls of our churches and government— holes that have been created by sin and idolatry.

Sadly, it seems that lately, there are more cracks in the wall than there are intercessors to help fill them. We see our nation crumbling around us— immorality, disease, drugs, and abortion. The Lord is pouring out His indignation upon the land.

Christians must take the battle to the front lines, to the enemy's own camp! We must cast down the strongholds of sin and unbelief in the citadels of power. Freemasonry is one of the greatest of these strongholds. This is partially because of its blatant idolatry, and its integral connection to politics.

Christians must begin spiritual warfare against the principalities of Freemasonry in their area, and for the land as a whole.

Ed Decker has addressed the problem of buildings dedicated by Masonic cornerstone layings in his amazing tract, *Freemasonry, Satan's Door to America.*[8] There he discloses the fact that even the streets of Washington, DC are laid out in Masonic patterns, including the square, compasses, and inverted pentagram. This is in addition to the huge Masonic idols like the Washington monument, the world's largest phallic symbol, or the Pentagon—a huge talisman of war!

Those concerned with the pro-life movement can also pray against the strongman of Masonry which looms over the black-robed priesthood of the Supreme Court, some of whom are undoubtedly Freemasons. If they serve gods like Ba'al and Molech, who demanded infant sacrifice, it is no wonder that they have ruled to permit the wholesale slaughter of babies.

On both the national and local levels, praying Christians must wrestle spiritually with Masonic principalities and powers (Ephesians 6:12). Baphomet, Jah-Bal-On, Baal, Hiram Abiff and Tubalcain are key demonic principalities which rule over the works of Masonry.

Authority must be taken over them in Jesus' name to bind their influence from all areas (Matthew 16:19, 18:18). Prayer groups or churches are better still because of the larger numbers of those praying in agreement (Matthew 18:20)!

It may be helpful to go to the government buildings of special concern and physically march around them, singing hymns about the blood and claiming them for Jesus—again, the more the merrier! As children of the living God, we can claim the promises of Deuteronomy 11:24, that:

> Every place whereon the soles of your feet
> shall tread shall be yours.

Find the cornerstones (they are almost always on the northeast corner), and take some oil, anoint it, claim it for Jesus and plead the blood over the entire building (school, post office, etc.). Pray that the strongholds of Masonry might be cast down, using the scriptures from II Corinthians 10:4-5 as your authority and reclaim your community for God!

Also, pray that the pastors of the churches in your community might be emboldened by the Holy Spirit to speak out publicly against the Lodge, and to deal with it in the local congregation. Pray for the guidance of the Holy Spirit as you do this, and He may bring things to mind in your own community which are specialized needs.

Finally, pray for the Masonic leadership in your town to repent. Find out the names of the Masters of the Lodges and place them specifically on your prayer list. Plead the blood of Jesus over them and if they are not Born Again pray that they would be led into a place where they might hear the gospel.

If they are saved, but deceived about Masonry, pray that the Spirit would convict them of their sin, and raise up many strong, knowledgeable and anointed voices in the community to speak out against the Ba'al worship of the Lodge.

If you can accomplish all this, and encourage your Christian friends to join with you in prayer, you will really put the devil and the Lodge to flight in your town. Then, by the grace of God, we can claim the precious promises of the Bible:

> If my people, which are called by my name, shall humble themselves, and pray, and seek my face, and turn from their wicked ways; then will I hear from heaven, and will forgive their sin, and will heal their land.
>
> II Chronicles 7:14

Appendix I

PRAYERS OF RENUNCIATION

If you are a Mason, you need to pray a prayer of repentance, renouncing the Lodge. You may wish to pray it with a friend or your pastor, although this is by no means essential. There is nothing "magical" in the words themselves, of course, as long as you get the thought across. Just pray with all your heart. You should say something like this:

"In the name of the Lord Jesus Christ and by the authority I possess as a believer in Him, I declare that I am redeemed out of the hand of the devil. Through the blood of Jesus, all my sins are forgiven. The blood of Jesus Christ, God's Son is cleansing me now from all sin. Through it, I am righteous, just as if I had never sinned.

"Through the blood of Jesus, I am sanctified, made holy, set apart for God—for I a member of a chosen generation, a royal priesthood, a holy nation, a peculiar people; that I should show forth Your praises, Lord God, Who has called me out of darkness into Your marvelous light (1 Pet. 2:9). My body is a temple for the Holy Spirit, redeemed, and cleansed by the blood of Jesus.

"I belong to the Lord Jesus Christ, body, soul and spirit. His blood protects me from all evil. In Jesus' name, I confess right now that I have been guilty of the sin of idolatry in the Masonic Lodge (I John 1:9). In agreement with the Lord, I call that involvement sin, and ask Him to completely remove it from my life and the lives of my family.

"In the name of Jesus, I rebuke any and all the lying and deceitful spirits of Freemasonry which may think that they still have a claim on me or my family (James 4:7). In Jesus' name, I renounce the spirits of Freemasonry, Ba'al, Jahbalon, Baphomet and Tubalcain and declare that you have no power over me any more, for I am bought and paid for by the blood of Jesus, shed on Calvary.

"I renounce any and all oaths made at the altar of Freemasonry in Jesus' holy name; and by the power of His shed blood I also break any generational sin and bondage which may be oppressing me through oaths made by my parents or ancestors (Exodus 20:5), and I nail all these things to the cross of Christ (Colossians 2:14).

I also break any and all power of the devil through these oaths over my own children or grand-children and command him to leave them alone, for they are under the blood of the Lamb of God! I also ask the Lord to cleanse any possible sin of illegit-imacy through the tenth generation with the blood of Jesus (Deuteronomy 23:2).

"Because of the blood of Jesus, Satan has no more power over me or my family, and no place in us. I renounce him and his hosts completely and declare them to be my enemies.

"Jesus said: 'These signs shall follow them that believe; in My name shall they cast out devils' (Mark 16:17). I am a believer, and in the name of Jesus I exercise my authority and expel all evil spirits. I command them to leave me now, according to the

Word of God and in the name of Jesus. "Forgetting those things which are behind, and reaching forth unto those things which are before, I press toward the mark for the prize of the high calling of God in Christ Jesus (Philippians 3:13-14). In Jesus' name. Amen."

PRAYER FOR BREAKING UNGODLY SOUL TIES

Where the "soul tie" issue is a concern, here is a sample prayer:

"Father, in the name of Jesus Christ, and by the authority I possess in Him as a believer, I ask you to cut any and all ungodly soul ties with my husband/wife (and anyone else with whom I have been intimate). I ask you to break any and all relationships established between us by Satan or his evil spirits, and ask that the blessed Holy Spirit would rebuild those relationships and covenant ties in accord with Your perfect will.

"I ask you to completely cleanse those relationships by the blood of the Lord Jesus Christ of all sin and demonic access. I declare to Satan in the name of Jesus Christ that he has no more access to my body, soul or spirit through those ties, and the doorway is shut and sealed by the blood of the Lamb!

"I also declare to Satan by the authority of the name of Jesus Christ that he has no power over my

270

children either. All access to my children through soul ties is hereby cut by the power of God, and I bring the full power of the Cross, the Blood, the Resurrection, and the Ascension of Jesus Christ against Satan's plans and schemes for me and my family. In Jesus' mighty name, Amen."

Though I hate to mention this unpleasant subject, in our ministry, we have found many cases of Masonic fathers, grandfathers or uncles who have abused children in the families sexually. This often comes from the spirit of lust which the sexual talismans of the Lodge provoke. In counseling with these people, it is often necessary for them to pray the above prayer to shut doorways which have been opened by their being raped incestuously as little children!

Sadly, there is little difference between the statistics for "Christian" versus non-Christian homes on child abuse, either sexual, emotional or physical. Masonry may be one of the primary reasons for that! Therefore, this is something that may need to be dealt with. Some adults have carried this horrid secret in silence their whole lives. Others have suppressed it to the extent they have forgotten it.

Obviously, there is no sin on the child's part, unless perhaps the child was a mid-to-late adolescent when the incest began! Even then, the much greater guilt lies at the door of the parent or relative who has trampled on the sacred position of trust the Lord accorded him over a precious young soul, and gravely misused the power and authority he wielded over the youngster.

Nevertheless, any sexual intimacy between child

and adult does create soul ties, as well as destroying the healthy love and assurance between parent and child. Therefore these things too need to be placed under the blood of Jesus. The child, (if old enough to understand), teen, or adult survivor of sexual abuse, needs to ask the Lord cleanse these ungodly ties with the relative, just as in the above prayer.

As horrible as all these things are, we must remember that Jesus' blood cleanses us from all sin and that He is waiting to forgive us for even the most awful transgressions. Give it to Him in prayer, and He will restore you!

FURTHER HELP OR COUNSEL

A Christian should always seek the help of his or her pastor first in these matters, assuming of course that the pastor is not a Mason himself. Sometimes, though, pastors (who cannot possibly be experts in everything) may not understand how to deal with the issues of Freemasonry or of generational bondage and deliverance. Also, it may help to talk with an ex-Mason who is now serving Jesus.

To that end, there are specialized evangelistic ministries within the Body of Christ designed to be a resource for those seeking help in escaping from the bondage of the Lodge. You may reach the author of this book in care of:

J. Edward Decker
Free the Masons Ministries
P.O. Box 1077,
Issaquah, WA. 98027
(206) 392-2077

In other parts of the U. S., you can also contact:

Mick Oxley*
In His Grip Ministries
Box 257-E—Rt#1,
Crescent City, FL. 32012
(904)649-5361

* A former Master Mason.

Rev. Harmon Taylor*
HRT Ministries
Box 43,
Redford, NY. 12148-0043

* A former Grand Chaplain, New York State.

Evangelist Jim Shaw*
Box 884,
Silver Springs, FL. 32688-0884

* A former 33° Mason.

Appendix II

LETTER OF RESIGNATION

This is a sample letter which one could send to resign from the Lodge. Although the exact wording is not important, it is ideal to strive for brevity, gentleness and a loving presentation of scriptural truth. Pray for the Lord to give you a truly loving approach. Long, preachy letters may not get read, even by the Lodge secretary. Of course, if you are a highly esteemed member of the Lodge, you may be able to get away with a longer letter.

Some have suggested that it is helpful to address the letter as done below, to the entire Lodge membership. Some Lodge bylaws demand that a letter so addressed be read in open Lodge. This would give you the maximum witnessing impact. Send a copy to your Lodge secretary and one to the "Master" of the Lodge, plus as many friends in the Lodge as you like:

> To the Master, Officers, and members
> of _____.

Dear Friends,

It is with regret that I submit my resignation from your Lodge, and from all Masonic bodies. I have taken this step in spite of the fact that I value very much the friendships and associations I have developed within Masonry over the years. However, One whom I love more than all of you has said, "He that loveth father or mother more than me is not worthy of me: and he that loveth son or daughter more than me is not worthy of me" (Matthew 10:37).

Those who call themselves Christians and believe in the name of Jesus are called to that kind of radical discipleship. If I am to be true to my Lord, Who loved me and gave Himself for me, then I must keep His commandments. If I am called to love Him more than my own family, how much more am I called to love him more than friends and fraternal brothers?

Unfortunately, in spite of the fraternal love I bear you all, I find that in following Jesus Christ, I must sever my ties with all forms of the Masonic fraternity. This is no reflection upon you, or upon the many good things that Masonry does. It is simply the fact that Masonry does not honor Jesus Christ as the Almighty God who came in the flesh to save us from our sins.

Our Lord said that, "He that is not with me is against me; and he that gathereth not with me scattereth abroad" (Matthew 12:30).

Masonry refuses to confess Jesus before men, and Jesus has warned that, "Whosoever therefore shall confess me before men, him will I confess also before my Father which is in heaven. But whosoever shall deny me before men, him will I also deny before my Father which is in heaven" (Matthew 10:32-33).

Anything which presents itself as a moral or religious institution and yet does not confess Christ as God is denying Jesus! I'm deeply sorry, brothers, but I cannot be a part of such a thing.

Masonry establishes itself as a complement to one's church, which sounds fine. Masonry has prayers, rituals and solemn ceremonies in which the authority of the Bible is invoked. The Lodge is not secular, but religious. But in my study of the Bible, it seems clear that Masonry asks us, as Masons, to do things contrary to Biblical teachings.

For example, Jesus commands us to "teach all nations" (Matthew 28:19-20) and to preach the gospel to everyone (Mark 16:15), yet Masonic etiquette forbids me as a Christian from sharing my Savior with my non-Christian Lodge brothers. I must stand by politely and watch them go to hell, "for there is none other name under heaven given among men, whereby we must be saved" (Acts 4:12). Who am I to obey, my Lord or my Lodge?

The very Bible which sits on the altar commands these things. It is one of the "three great lights" of Masonry, and yet you ignore its teachings for the sake of harmony!

Jesus also commands us not to swear oaths (Matthew 5:34-37) and His sovereign command is echoed by James (5:12). Yet the taking of solemn oaths is at the very heart of Masonic degree work. My brothers, these things ought not to be! Finally, the apostle Paul commands that believers in Jesus Christ be "...not unequally yoked together with unbelievers: for what fellowship hath righteousness with unrighteousness? and what communion hath light with darkness?" (II Corinthians 6:14).

I take my Masonic obligations very seriously, and

for that very reason, I realize the cable-tow was a powerful yoke which bound me to many Masons who do not worship the true God or His Son Jesus, however sincere they may be in their devotions. I must take the commands of my God more seriously. To remain in Masonry is to compromise my fellowship with Jesus Who died for me.

I realize that whatever benefits Masonry has, they cannot possibly be compared with the joys of full friendship with the Almighty Lord of the Universe Who died that I might live—and Who loves all Masons, as I do, even though they allow Him no place in their ceremonies.

You, my brothers, must decide for yourselves what you will do with the god of Masonry, but "as for me and my house, we will serve the LORD." (Joshua 24:15)

God bless you as you seek His will in this important matter.

In Christian love,

Footnotes

Chapter 1

1. Duncan, M. C., *Duncan's Masonic Ritual and Monitor,* David McKay Company, Inc., New York, n.d., p. 36.

Chapter 2

1. Duncan, Malcolm C., *Duncan's Masonic Ritual and Monitor,* p. 31.
2. Coil, Henry Wilson, *Coil's Masonic Encyclopedia,* Macoy Publishing, Richmond, VA, 1961, p. 51.
3. Ibid., p. 51.
4. Duncan, Malcolm C., *Duncan's Masonic Ritual and Monitor,* p. 95,134.
5. Ibid., p. 95.
6. Ibid., p. 12.
7. Hall, Manly P., *The Phoenix,* The Philosophical Research Society, Los Angeles, 1975, p. 37.
8. Pike, Albert, *Morals and Dogma of the Ancient and Accepted Scottish Rite of Freemasonry,* Supreme Council of the Thirty-Third Degree, Charleston, 1950, p. 213.
9. Ibid p. 219.
10. Pike, Albert, *Morals and Dogma...* p. 213.
11. Ibid., p. 718.
12. Mackey, Albert, *Mackey's Revised Encyclopedia of Freemasonry,* Macoy Publishing, Richmond, VA., 1966, p. 618.
13. Ankerberg, John; Weldon, John, *The Secret Teachings of the Masonic Lodge,* Moody Press, Chicago, 1990, p. 16-17.

Chapter 3

1. Duncan, M. C., *Duncan's Masonic Ritual and Monitor,* p. 10.
2. Ibid., p. 95.
3. Ibid., p. 96.
4. Ibid., p. 143

Chapter 4

1. Duncan, Malcolm C., *Duncan's Masonic Ritual and Monitor,* p. 30.
2. Mackey, Albert, *Mackey's Revised Encyclopedia of Freemasonry,* pp 409-410.
3. Coil, Henry Wilson, *Coil's Masonic Encyclopedia,* pp 516-517.
4. Schnoebelen, William, *Wicca: Satan's Little White Lie,* Chick Publications, Chino, CA, 1990.
5. Coil, Henry Wilson, *Coil's Masonic Encyclopedia,* pp 516-517.
6. Pike, Albert, *Morals and Dogma...* p. 226.

7. Ibid., p. 525.
8. Hall, Manly P., *The Lost Keys of Freemasonry,* Macoy Publishing Richmond, VA, 1976, p. 65.

Chapter 5
1. Duncan, Malcolm C., *Duncan's Masonic Ritual and Monitor,* p. 36.
2. Ibid., pp 118, 120.
3. Ibid., p. 249.
4. Flexner, Stuart Berg, *The Random House Dictionary of the English Language,* Hauck, L. C. Random House, New York, 1987, p. 1352.
5. Blanchard, J., *Scottish Rite Masonry Illustrated, The Complete Ritual,* Ezra Cook Publications, Chicago, 1974, pp 453-457.
6. Pike, Albert, *Morals and Dogma...* pp 321.
7. Duncan, p. 35.
8. Pike, p. 102.
9. de LaRive, A. C., *La Femme et l'Enfant dans la Franc,* Maçonnerie Universelle, Paris, 1889, p. 588.
10. Hall, Manly P., *The Lost Keys of Freemasonry,* p. 48.

Chapter 6
1. Duncan, Malcolm C., *Duncan's Masonic Ritual and Monitor,* p. 122.
2. Ibid., p. 221.
3. Ibid., pp 155, 190, 208, 230.
4. *Illustrated Ritual of the Six Degrees of the Council and Commandery,* Charles Powner, Co., Chicago, 1975, p. 202.
5. Ibid., p. 212.
6. Ibid.
7. Ibid., p. 217.
8. Ibid., pp 200, 224, for example.
9. Ibid., p. 217.
10. Ibid., p. 215.
11. Ibid., pp 227-28.
12. Ibid., p. 211.
13. Pike, Albert, *Morals and Dogma...* p. 525.
14. Clausen, Henry C., *Practice and Procedure for the Scottish Rite;* Supreme Council of the Thirty-Third degree of the Ancient and Accepted Scottish Rite of Freemasonry, Washington, DC, 1981, pp 75-77.
15. Pike, p. 539.
16. Ibid., p. 226.
17. Clymer, R. Swineburne, *The Mysticism of Masonry,* 1900, p. 47.
18. Buck, J.D., *Symbolism or Mystic Masonry,* 1925, p. 57.

Chapter 7

1. Blanchard, J., *Scottish Rite Masonry Illustrated, The Complete Ritual,* Ezra Cook Publications, Chicago, 1974, II, p. 47.
2. Pike, Albert, *Morals and Dogma...* pp 167.
3. Mackey, Albert, *Mackey's Revised Encyclopedia of Freemasonry,* Macoy Publishing, Richmond, VA, 1966, p. 192.
4. Duncan, Malcolm C., *Duncan's Masonic Ritual and Monitor,* p. 29.
5. Ibid., p. 33.
6. Robbins, Russell Hope, *The Encyclopedia of Witchcraft and Demonology,* Crown Publishers, New York, 1959, p. 420.

Chapter 8

1. Duncan, Malcolm C., *Duncan's Masonic Ritual and Monitor,* p. 34-35.
2. Ibid., p. 95.
3. Ibid., p. 230.
4. Ibid., p. 229.

Chapter 9

1. Shephard, Leslie A., *Encyclopedia of Occultism and Parapsychology,* Avon Books, New York, 1980. 2, p. 552.
2. LaVey, Anton Szandor, *The Satanic Bible,* Avon Books, New York, 1969, front cover.
3. Godwin, Jeff, *The Devil's Disciples,* Chick Publications, Chino, CA, 1987, p. 172.
4. Godwin, Jeff, *Dancing With Demons,* Chick Publications, Chino, CA, 1989, p. 181.
5. Schnoebelen, William J., *The F.A.T.A.L. Flaw,* Saints Alive in Jesus, Issaquah, WA, 1987.
6. Michell, John, *The City of Revelation,* Ballentine Books, New York, 1972, p. 4.
7. Grant, Kenneth, *The Magical Revival,* Samuel Weiser, New York, 1973, pp 43-44, 65.
8. Ronayne, Edmond, *Blue Lodge and Chapter,* Ezra A. Cook, Chicago, n.d., pp 87-88.
9. Pike, Albert, *Morals and Dogma...* pp 14-15.
10. Lyons, Arthur, *Satan Wants You,* Mysterious Press, 1988, pp 126-27.
11. Aquino, Michael A., *The Wewelsburg Working,* October 19, 1984.

Chapter 11

1. Duncan, Malcom C., *Duncan's Ritual Monitor,* pp 158-183.
2. Ibid., pp 158, 172.
3. Hislop, Rev. Alexander, *The Two Babylons,* Loizeaux Bros., Neptune, NJ, 1959, p. 204.

4. *Illustrated Ritual of the Six Degrees...* p. 232.

5. Hislop, pp 202-205.

6. Ibid., p. 204.

7. Regardie, Israel, *The Golden Dawn,* Llewellyn Publications, Minneapolis, 1971, vol.III, p. 49.

8. Whittemore, Carrol E., *Symbols of the Christian Church,* Abingdon Press, Nashville, 1959, p. 12.

9. Crowley, Aleister, *The Equinox, Vol.3, No.1,* Samuel Weiser, New York, 1973, pp 226, 248.

10. Jones, Alexander (gen. ed.), *The Jerusalem Bible,* Doubleday & Co., Garden City, NY, 1970, frontispiece.

11. Flexner, Stuart B., *The Random House Dictionary of the English Language,* Second Edition, Random House, NY, 1987, p. 475.

12. Pike, *Morals and Dogma...* p. 291.

13. Flexner, p. 1455.

14. Pike, p. 14, 792.

15. Dualism is the belief in two equal but opposite forces in the universe, good and evil. Contrary to popular belief, the devil is not the opposite of God, because that would mean the devil and God are equals. They are not! God is so far above the devil in power that it cannot be comprehended.

16. Hall, Manly P., *The Secret Teachings of All Ages,* p. cc.

17. Ibid.

18. Crowley, Aleister, *777,* privately printed by the O.T.O., 1907, p. 13.

19. Schnoebelen & Spencer, *Mormonism's Temple of Doom,* pp 20-23.

20. Duncan, op.cit., p. 97.

21. Ibid. p. 92.

22. Ibid., p. 143.

23. Ibid., p. 95.

Chapter 12

1. Pike, Albert, *Morals and Dogmas...* p. 105.

2. Ibid., pp 744-45.

3. Ibid. p. 732.

4. Mackey, Albert, *Mackey's Revised Encyclopedia of Freemasonry,* p. 133.

5. Coil, Henry Wilson, *Coil's Masonic Encyclopedia,* p. 520.

6. Street, Oliver Day, *Symbolism of the the Three Degrees;* Masonic Service Association, Washington, DC, 1924, pp 44-45.

7. Ibid., pp 46-47.

8. Duncan, op.cit., p. 129.

9. Roberts, Allen E., *The Craft and Its Symbols; Opening the Door to Masonic Symbolism,* Macoy Publishing, Richmond, VA, 1974) p. 76.

10. Duncan, p. 132.
11. Ibid., p. 39.
12. Allen, Raymond Lee et.al, *Tennessee Craftsmen or Masonic Textbook,* Tennessee Grand Lodge, Nashville, 1983, p. 17.

Chapter 13
1. Pike, p. 819.
2. Duncan, p. 137.
3. Pike, pp 104-105.
4. Duncan, p. 36.
5. Mackey, Albert, *The Manual of the Lodge,* Clark Maynard Co., New York, 1870, p. 156.
6. Mackey, Albert, *Mackey's Revised Encyclopedia of Freemasonry,* Macoy Publishing, Richmond, VA, 1966, p. 560.
7. Schnoebelen, William; Spencer, James R., *Mormonism's Temple of Doom,* Triple J Publications, Boise, ID, 1987, p. 13.
8. Ronayne, Edmond, *Handbook of Freemasonry,* Ezra Cook Publications, Chicago, IL, 1976, p. 177.
9. Mackey, Albert, *Mackey's Revised Encyclopedia of Freemasonry,* p. 513.
10. Alexander, David; Alexander, Patricia, *Eerdmans' Handbook to the Bible,* Eerdmann's, Grand Rapids, MI 1984, p. 135.
11. Hislop, Rev. Alexander, *The Two Babylons,* pp 28-38.
12. Schnoebelen, William, *Wicca: Satan's Little White Lie,* Chick Publications, Chino, CA., 1990, pp 169-175.
13. Chick, Jack T., *Angel of Light,* The Crusaders, Vol. 9, Chick Publications, Chino, CA, pp 13-16.
14. Duncan, pp 102-121.

Chapter 14
1. Pike, Albert, *Morals and Dogmas...* p. 839.
2. Flexner, Stuart Berg, *The Random House Dictionary of the English Language,* Random House, New York, 1987, p. 1272.
3. Cavendish, Richard, *Man, Myth and Magic,* Marshall Cavendish Corporation, New York, 1970,14. p. 1925.
4. Steinmetz, George H., *The Royal Arch-Its Hidden Meaning,* Macoy Publishing, Richmond, VA, 1946, pp 51-52.
5. Frazer, Sir James George, *The Golden Bough,* MacMillan & Co., New York, 1960, p. 459.
6. Powell, Neil, *Alchemy, the Ancient Science,* Wilson, C., Aldus Books, Ltd., London, 1976, p. 30.
7. Eliot, Alexander; Campbell, Joseph, *Myths,* McGraw-Hill Books, New York, 1976, p. 258.

^ Grant, Kenneth, *Aleister Crowley and the Hidden God*, Samuel Weiser, New York, 1974, p. 151.
`. pp 151-152.
J. Eliot, p. 258.
11. Schnoebelen, William J.; Spencer, James R., *Whited Sepulchers, the Hidden Language of the Mormon Temples*, Triple J Publications, Boise, ID,1990, p 21.
12. Duncan, Malcolm C., *Duncan's Masonic Ritual and Monitor*, David McKay Company, Inc., New York, n.d., pp 118, 120.
13. Ibid., p. 125.
14. Flexner, p. 305.
15. Pike, p. 401
16. Ibid., p. 402
17. Ibid., p. 22.
18. Mackey, Albert, *Mackey's Revised Encyclopedia of Freemasonry*, p. 497
19. Pike, p. 839.

Chapter 15
1. DaCosta, Hippolyto Joseph, *The Dionysian Artificers*, Philosophical Research Society, Los Angeles, 1964 (1820).
2. DaCosta, p. xiv.
3. Flexner, Stuart Berg, *The Random House Dictionary of the English Language*, Random House, New York, 1987, p. 557.
4. Duncan, Malcolm C., *Duncan's Masonic Ritual and Monitor*, p. 129.
5. Edwards, Paul, *The Encyclopedia of Philosophy*, MacMillan, New York, 1972, vol. VII, p. 37.
6. Fargis, Paul, *The New York Public Library Desk Reference*. Webster's New World, New York, 1989, p. 217.
7. Furnivall, F.J., *The Compact Edition of the Oxford English Dictionary*, Oxford University Press, Oxford, 1979, 2. p. 2762.
8. LaVey, Anton Szandor, *The Satanic Rituals*, University Books, Secaucus, N.J., 1972, pp 151-155.
9. Furnivall, F.J., 1. p. 125.
10. Baigent, Michael, Leigh Richard, *The Temple and the Lodge*, Arcade Publishing, New York, 1989, p. 45.
11. Ibid., p. 42.
12. Seward, D., *The Monks of War*, St. Alban's Press, 1974, p. 37.
13. Baigent, Leigh, p. 43.
14. Ibid., p. 51.
15. Ibid., p. 53.
16. Valiente, p. 102.
17. Furnivall, F.J., 1. p. 165.

283

18. Valiente, p. 102.
19. Wright, Thomas, in *"Essays on the Worship of the Generative Powers during the Middle Ages of Western Europe"*, Knight, R.P., London, 1865
20. Grant, Kenneth, *The Magical Revival*, Samuel Weiser, New York, 1973, p. 71.
21. Shephard, Leslie A., *Encyclopedia of Occultism and Parapsychology.* Avon Books, New York, 1980, 2. p. 632.
22. Grant, Kenneth, *The Magical Revival* p. 72.
23. Fortune, Dion, *Psychic Self-Defense*, Samuel Weiser, New York, 1972, p. 149.
24. Hall, Manley P., *The Secret Teachings of All Ages, An Encyclopedia Outline of Masonic, Hermetic, Qabbalistic and Rosicrucian Symbolical Philosophy,* The Philosophical Research Society, Los Angeles, 1978, p. ci.
25. Baigent, Leigh, p. 78.
26. Gantz, Jeffrey, *Mabinogion,* Penguin, London, 1976.
27. Valiente, pp 182-183.
28. Ibid., p. 60.
29. Hislop, Rev. Alexander, *The Two Babylons,* pp 33-35.
30. Valiente, p. 160.

Chapter 16
1. Haywood, H. L., *The Great Teachings of Masonry,* Macoy, Richmond, VA, 1971, p. 94.
2. Baigent, Michael; Leigh, Richard, *The Temple and the Lodge,* Arcade Publishing, New York, 1989, pp 10-11.
3. Ibid., p. 106.
4. Ibid., p. 92.
5. Valiente, Doreen, *An ABC of Witchcraft,* Phoenix, Custer, WA, 1988, p. 159.
6. Farrar, Janet; Farrar, Stewart, *Eight Sabbats for Witches,* Robert Hale, London, 1981, see photo plate #15.
7. Valiente, p. 159.
8. McIntosh, Christopher, *The Rosy Cross Unveiled,* The Aquarian Press, Ltd., Wellingborough, Northamptonshire, 1980, p. 19.
9. Daraul, Arkon, *A History of Secret Societies,* Citadel Press, Secaucus, NJ, 1961, pp 191-192.
10. Hall, Manly P., *The Secret Teachings of All Ages...* pp cxxxvii.
11. Ibid., p. cxxxviii.
12. Daraul, p. 192.
13. Ibid.
14. Hall, *Secret Teachings...* p. clxi.

15. Ibid.
16. Powell, Neil, *Alchemy, the Ancient Science,* Aldus Books, Ltd., London, 1976, pp 8-11.
17. Pike, Albert, *Morals and Dogma...* p. 731.
18. Baigent, Leigh, pp 118-19.

Chapter 17
1. Hall, Manly P., *The Secret Teachings of All Ages...* p. cxxxix.
2. Yates, Francis, *The Rosicrucian Enlightenment,* St. Alban's Press, 1975, p. 226.
3. LaVey, Anton, *The Satanic Bible,* Avon Books, New York, 1969, pp 155-156.
4. Baigent, Michael; Leigh, Richard, *The Temple and the Lodge,* Arcade Publishing, New York, 1989, p. 155.
5. Pick, F.L.; Knight, G.N., *The Pocket History of Freemasonry,* London, 1983, p. 45.
6. Valiente, Doreen, *An ABC of Witchcraft,* Phoenix, Custer, WA, 1988, p. 91.
7. Ibid., 1988, p. 181.
8. Ibid., 1988, p. 318.
9. Canseliet, Eugene, *Fulcanelli: Master Alchemist, 'Le Mystere des Cathedrales,'* Neville Spearman, London, 1971.
10. Waite, Arthur Edward, *A New Encyclopedia of Freemasonry,* Weathervane Books, New York, 1970, p. 326.
11. Coil, Henry Wilson, *Freemasonry through Six Centuries,* Macoy Publishing, Richmond, VA, 1967, p. 131.
12. Waite, p. 330.
13. Ibid., 1970, p. 66.
14. Farrar, Janet; Farrar, Stewart, *Eight Sabbats for Witches,* pp 80-81.
15. Daraul, Arkon, *A History of Secret Societies,* Citadel, Secaucus, NJ, 1961, p. 220.
16. Heckethorn, Charles William, *The Secret Societies of All Ages and Countries,* University Books, New Hyde Park, NY, 1966, pp 305-306.
17. Waite, p. 387.
18. Schnoebelen, William, *Wicca: Satan's Little White Lie,* Chick Publications, Chino, CA, 1990, p. 116-117.
19. Webster, Nesta H., *Secret Societies and Subversive Movements,* Britons Publishing Co., London, 1964, p. 235.

Chapter 18
1. Waite, Arthur Edward, *A New Encyclopedia of Freemasonry,* Weathervane Books, New York, 1970, p. 278.
2. Margiotta, Domenico Adrianno Lemmi, 1894.

3. de LaRive, A.C., *La Femme et l'Enfant dans la Franç,* Maconnerie Universelle, Paris, 1889, p. 588.
4. Heckethorn, Charles William, *The Secret Societies of All Ages and Countries,* University Books, New Hyde Park, NY, 1966, p. 279.
5. Duncan, Malcolm C., *Duncan's Masonic Ritual and Monitor,* p. 29.
6. Epperson, A. Ralph, *The Unseen Hand: An Introduction to the Conspiratorial View of History,* Publius Press, Tucson, AZ, 1985, p. 223.
7. Queensborough, Lady (Edith Starr Miller) *Occult Theocracy,* Christian Book Club of America, Los Angeles, 1933, p. 217.

Chapter 19
1. Schnoebelen, William J., Spencer, James R., *Whited Sepulchers, the Hidden Language of the Mormon Temples,* Triple J Publications, Boise, ID, 1990, pp 44-50, citing Anton LaVey's newsletter, "The Cloven Hoof," vol. VIII, #6.

Chapter 20
1. Duncan, Malcolm C., *Duncan's Masonic Ritual Monitor,* p. 15
2. Ibid., p. 52
3. Ibid., p. 36
4. Ibid., p. 65
5. Ibid., p. 102

Chapter 21
1. Schnoebelen, William J., *Wicca: Satan's Little White Lie,* pp 116-122, 166-174.
2. Ibid., pp 104-111.

Chapter 23
1. Knollenberg, Bernard, *George Washington, The Virginia Period,* Duke University Press, Durham, NC, 1964, p. 10.
2. Ibid., p. 145.
3. *Congressional Record—Senate,* September 9, 1987, S 11868-70.
4. Ibid., p. 11870.
5. Dolinko, Cary N. et.al., *Freemasonry,* May 19, 1987, Tara Center
6. Ibid.
7. Creme, Benjamin, *The Reappearance of the Christ,* pp 84,87.
8. Decker, J. Edward, *Freemasonry, Satan's Door To America,* Free The Masons Ministries, Issaquah, WA, 1988.